An Introduction to Coastal Engineering

Michael Isaacson, P. Eng.
Professor Emeritus of Civil Engineering
The University of British Columbia, Canada

Published by John Wiley & Sons, Inc., Hoboken, New Jersey.
Published simultaneously in Canada.

For general information on our other products and services or for technical support, please contact our Customer Care Department within the United States at (800) 762-2974, outside the United States at (317) 572-3993 or fax (317) 572-4002.

Wiley also publishes its books in a variety of electronic formats. Some content that appears in print may not be available in electronic formats. For more information about Wiley products, visit our web site at www.wiley.com.

Library of Congress Cataloging-in-Publication Data Applied for:

Hardback ISBN: 9781394257140

Cover Design: Wiley
Cover Image: © Raul/Adobe Stock

Set in 9.5/12.5pt STIXTwoText by Straive, Chennai, India

SKY10089465_102924

To Sharon, Ben, Abby, Ian, and Jordanna.

Contents

About the Author

Dr. Michael Isaacson is Professor Emeritus of Civil Engineering at the University of British Columbia (UBC). He received his degrees from the University of Cambridge and has been active throughout his career in teaching, research, university service, professional service, and engineering practice.

Dr. Isaacson's teaching contributions have included the delivery of a course in coastal engineering over several decades. His research interests relate to coasal engineering and he is the author/co-author of over 200 technical papers in his field. His service contributions have included roles as Head of Civil Engineering and Dean of Applied Science at UBC, journal editorships, and leadership roles on national and international professional committees and professional associations. Dr. Isaacson is the recipient of many national and international career achievement awards, best paper awards, and professional service awards.

Dr. Isaacson is a professional engineer and throughout his career has contributed to a wide range of local, national, and international engineering projects. These have included projects relating to design wave and wave load predictions, the motions of floating structures, assessments of tsunami and hurricane impacts, sedimentation studies, sea level rise and coastal flooding assessments, laboratory model testing, and the design of marinas and coastal infrastructure. Dr. Isaacson remains active in coastal engineering practice.

Preface

Coastal engineering relates to the solution of engineering problems in the coastal environment. It concerns, for example, the design of coastal structures such as breakwaters and seawalls, the design of harbors and marinas, understanding and addressing the consequences of sea level rise and coastal flooding, the assessment and control of sediment erosion and accretion, the design of shoreline protection schemes, and the assessment and controlled discharge of pollutants into the ocean environment.

This text stems from the author's delivery of a combined undergraduate/graduate course in coastal engineering at the University of British Columbia over many years and from his engagement in coastal engineering practice over many years. It is intended to serve simultaneously as an introductory text directed to final-year undergraduate students, an advanced level text directed to graduate students, and a more general guide for practicing engineers engaged in coastal engineering projects. Some prior knowledge and understanding of applied mathematics and fluid mechanics is assumed.

The text includes the fundamental development of relevant concepts, a compendium of formulations, a series of illustrations and photographs (a number of the latter are drawn from British Columbia, reflecting the author's experiences), a set of worked examples that illustrate basic calculations or the application of available spreadsheet solutions, and a set of problems and/or written assignments at the end of each chapter. The latter are written in generic form, and may be adapted by an instructor with respect to particular locations or circumstances.

The text contains three appendices. Appendix A, which may be downloaded as a spreadsheet from the text's companion site, provides a set of reference solutions relevant to various formulations that are outlined in some chapters. Appendix B provides a list of the notation that is used and Appendix C provides values of physical constants that are used most frequently in coastal engineering. The text is based on the SI system of units.

In order that this text may serve simultaneously as an undergraduate and a graduate text, portions of the text considered to be at a more advanced level are contained within the symbols ■➔ and ←■, while problems or parts of problems considered to be at a more advanced level are preceded by the symbol ■.

This text is intended to serve as an introductory text and accordingly it does not encompass more advanced aspects of coastal engineering coverage. Thus, it does not provide detailed derivations of various formulae, but rather focuses on more general descriptions of how these are developed. While it makes extensive reference to numerical models, it does not provide sufficient information for the reader to be able to develop such models, which would rely on a thorough understanding of the underlying phenomena and of computational methods, nor to be able to utilize such models, which typically depends on users guides and specific training. Instead, the focus is on the development and use of spreadsheets to provide more fundamental solutions to coastal engineering problems. Finally, the text does not include a comprehensive list of references, but rather relies largely on references to other texts and manuals, sometimes implied, while more specific studies are only referenced as may be necessary.

Coastal engineering practice relies on access to and a reliance on various data sources, including wind records, tide records, hydrographic charts, bathymetry, and sea level rise projections. The format and availability of such sources vary worldwide. Where appropriate, commentary on access to such sources are provided, with a focus on data sources that are relevant to the United States and to Canada. If required, it is hoped that the reader may be able to find and utilize equivalent data sources in other countries as may be relevant.

The printed version of this text is in black-and-white, while the online version is in color. The reader is referred to the online version in order to view in color many of the figures including all those comprising photographs and graphics.

This work is an outcome of the author's engagement over many years in coastal engineering research, teaching, and practice. Recognizing this, the author wishes to gratefully acknowledge the related collaborations, discussions, and contributions provided by his colleagues in the engineering profession and by faculty members and former students at the University of British Columbia and elsewhere.

The author hopes that the text contributes to the reader's understanding of coastal engineering and ability to develop solutions to coastal engineering problems.

September 2024

Michael Isaacson
Vancouver, BC, Canada

About the Companion Website

This book is accompanied by a companion website:

www.wiley.com/go/coastalengineering

The website includes links that enable readers to download a spreadsheet *Problem Data* that contains data referred to in selected problems and a second spreadsheet *Appendix A* that contains a set of reference solutions. An additional link enables instructors to download relevant instructor resource materials.

1

Introduction

1.1 Scope of Coastal Engineering

Coastal engineering relates to the solution of engineering problems in the coastal environment. It concerns, for example: the design of coastal structures such as breakwaters and seawalls, the design of harbors and marinas, understanding and addressing the consequences of sea level rise and coastal flooding, the assessment and control of sediment erosion and accretion, the design of shoreline protection schemes, and the assessment and controlled discharge of pollutants in the ocean environment.

Coastal engineering is a branch of civil engineering. Neighboring civil engineering subdisciplines that may be relevant to a coastal engineering project include hydraulic engineering, geotechnical engineering, structural engineering, and earthquake engineering. Neighboring disciplines include ocean engineering that relates to engineering projects in the ocean, uninfluenced by proximity to a coastline and often associated with offshore oil and gas recovery; naval architecture that relates to the design and operation of ships and marine vessels; and oceanography that relates to the scientific study of the oceans. Some subdisciplines of the sciences, such as nearshore oceanography and coastal geology, overlap with aspects of coastal engineering.

Several decades ago, Weigel (1964) and Ippen (1966) developed what may be regarded as foundational texts in coastal engineering. Subsequently, several texts on coastal engineering that have been available and widely used include Sorensen (2006), Sawaragi (2011), Reeve et al. (2018), and Kamphuis (2020), with each one providing different areas of emphasis. Many other texts focus on particular aspects of coastal engineering, such as marina design, port engineering, and coastal processes, or on topics within related disciplines such as ocean engineering and hydraulic engineering. In addition, various design manuals and guides that are relied upon in coastal engineering practice include the Coastal Engineering Manual (2002) and its predecessor the Shore Protection Manual (1984), the Rock Manual (2007), the Federal Emergency Management Agency's (FEMA) Coastal Construction Manual (2011), and the EurOtop Manual (2018).

1.2 Outline of Book

Since ocean waves are usually the primary environmental consideration in coastal engineering, it is customary that a major part of a coastal engineering text relates to a description of waves. In this context, the chapter titles of the book are as follows:

An Introduction to Coastal Engineering, First Edition. Michael Isaacson.

Companion website: www.wiley.com/go/coastalengineering

1. Introduction
2. Regular Waves
3. Wave Transformations
4. Random Waves
5. Winds
6. Wave Predictions
7. Long Waves, Water Levels, and Currents
8. Coastal Structures
9. Coastal Processes
10. Mixing Processes
11. Design of Coastal Infrastructure
12. Coastal Modeling

More specifically, summaries of these chapters are given below.

Chapter 2, *Regular Waves*, describes the treatment of regular waves – that is, periodic waves that propagate over a horizontal seabed without a change in form. This chapter focuses on the development and application of linear wave theory.

Chapter 3, *Wave Transformations*, considers the transformation of waves associated with changes in water depth and with obstacles in the flow. Thus, the chapter treats wave shoaling, wave refraction, wave diffraction, wave reflection including standing waves, wave transmission, wave attenuation, wave breaking, and wave runup at a shoreline.

Chapter 4, *Random Waves*, recognizes the random nature of waves and distinguishes between the short-term variability of individual waves over a few hours and the long-term variability of different storms over several years, typically leading to design wave conditions with a specified return period.

Chapter 5, *Winds*, which serves as a prelude to Chapter 6, gives attention to descriptions of a wind climate, approaches to accessing and analyzing wind data needed for wave hindcasting, and a description of the wind field in a hurricane.

Chapter 6, *Wave Predictions*, describes the prediction of waves, with a focus on a simplified approach to wave hindcasting that provides estimates of wave conditions on the basis of available wind data. Other forms of wave prediction that are mentioned include ship waves and laboratory-generated waves.

Chapter 7, *Long Waves, Water Levels, and Currents*, outlines long waves, which include tides and tsunamis, and coastal flooding water levels and their components, which include storm surge and sea level rise. This chapter also indicates the impacts of climate change on coastal engineering practice and concludes with a summary of the kinds of currents that may be encountered.

Chapter 8, *Coastal Structures*, summarizes various categories of coastal structure and approaches to estimating wave loads on structures and wave interactions with structures. The kinds of structures considered include seawalls, rubble-mound structures, slender-member structures, large structures, and floating structures. This chapter also provides descriptions of the analysis of floating breakwaters and floating bridges and an outline of loads other than wave loads that may act on a coastal structure. The chapter concludes with a description of ocean-related renewable energy infrastructure.

Chapter 9, *Coastal Processes*, begins with an outline of the variety of coastal forms that may be encountered. It then describes in turn coastal sediments, the conditions for the onset of sediment movement under currents and waves, the movement of sediments near shorelines including sediment transport along beaches, bluff erosion, and scour, mitigation measures for addressing unwanted sediment erosion or accretion, approaches to shoreline protection, including reliance on nature-based methods, coastal restoration, and, finally, an introduction to coastal management.

Chapter 10, *Mixing Processes*, which is a primary topic of environmental fluid mechanics, describes fundamental solutions to the advection–diffusion equation and summarizes related topics that include stratified flows, mixing in estuaries, and jets and plumes.

Chapter 11, *Design of Coastal Infrastructure*, provides an introduction to the design process and approaches to accounting for uncertainty and outlines selected design tools including probability of failure analyses, risk assessment and management, permitting and approval requirements, and decision-making in design. It then summarizes selected design considerations with respect to coastal structures, including their modes of failure, design criteria, and aspects of detailed design. Finally, the chapter discusses the design of harbors and marinas, with attention given to available criteria for acceptable wave climate in marinas and to navigational considerations.

Chapter 12, *Coastal Modeling*, summarizes in a general way numerical modeling associated with various coastal engineering phenomena. It then outlines the underlying principles of model laws, describes different kinds of laboratory models used in coastal engineering, and provides descriptions of laboratory facilities, instrumentation, and measurements. The chapter concludes with a summary of field measurements relevant to coastal engineering.

1.3 Example Projects

In order to provide an appreciation of the breadth of coastal engineering practice, four generic example projects are now presented, along with an indication of the range of issues that may require consideration within each of these. (Any specific terminology that is used here is defined subsequently within the text.)

1.3.1 Coastal Flooding

The first generic project relates to an assessment of coastal flooding along a shoreline and the design of flood protection infrastructure. Figure 1.1 shows a coastal dike in Richmond, BC, used here as a basis for identifying key design parameters and associated issues.

The dike design entails the selection of the dike's crest elevation and its sectional properties (e.g. crest width, seaward and landward slopes, rock size, vegetation, ...) and a consideration of drainage and pump systems. The design is influenced by maximum water levels due to a combination of tides, sea level rise, storm surge and waves, and/or, if relevant, river flooding or tsunami flooding; a consideration of the probability and extent of any flooding in the context of uncertainty, risk, return period, and design life; wave runup and overtopping; and, finally, potential habitat enhancements, climate change impacts, permitting and approval requirements, and land use requirements. While Figure 1.1 indicates the case of a coastal dike, related coastal flooding projects may involve a seawall or other coastal defense along a shoreline, and a consideration of hurricane-prone areas for which storm surge is a major issue.

Figure 1.1 Coastal dike example project.

1.3.2 Coastal Structure Design

A second generic example relates to the design of different types of coastal structure. Figure 1.2 illustrates four types of structure: a seawall, rubble-mound shoreline protection, a piled pier, and a floating structure, with the figure used as a basis for identifying key design parameters and associated issues.

Common to all these are assessments of tides, water levels and wave climate, and an identification of return periods and suitable levels of uncertainty and risk. In addition, the design of seawalls entails a determination of wave loads, wave runup, and wave overtopping. Rubble-mound shoreline protection entails a determination of wave overtopping and a consideration of rubble-mound stability, which depends in part on the rubble-mound slope, rock type, and rock size. A piled structure entails a determination of wave loads on piles and the deck elevation. Finally, a floating structure usually entails an assessment of structure motions and mooring system and anchor design.

1.3.3 Sediment Transport

A third generic example project relates to an assessment of sediment transport along a beach or shoreline, along with any mitigation measures that may be undertaken. Figure 1.3 shows the shoreline at Spanish Banks in Vancouver, BC, used here as a basis for identifying key design parameters and associated issues.

Key considerations include sediment properties, the onshore–offshore transport of sediment during and between storms, the longshore transport of sediment (parallel to the shore), an assessment of beach slope and profile, and beach protection measures that may be introduced, such as groins, offshore breakwaters, and rock mounds. The design may be influenced by water levels, a consideration of uncertainty,

Figure 1.2 Types of coastal structure projects. (a) Seawall, (b) rubble-mound shoreline protection, (c) piled pier, (d) floating structure. *Source:* ShoreZone/CC BY 3.0.

Figure 1.3 Coastal sediment transport example project. *Source:* ShoreZone/CC BY 3.0.

Figure 1.4 Marina design example project. *Source:* Google Earth.

risk, return period, and design life, and, finally, potential habitat enhancement, climate change impacts, and permitting and approval requirements.

1.3.4 Marina Design

A fourth generic example project relates to the design of a marina. Figure 1.4 shows an example marina at Gibsons, BC, used here as a basis for identifying key design parameters and associated issues.

Key aspects of the design include the breakwater layout intended to achieve acceptable wave conditions within the marina, and the breakwater's sectional design (i.e. crest elevation, crest width, slope, and rock type, size, and placement), so as to assure the effectiveness and stability of the breakwater. The design is influenced by design wave conditions approaching the marina; water levels (due to a combination of tides, sea level rise, storm surge, and waves); a consideration of uncertainty, risk, return period, and design life; and the infrastructure within the marina, including slips, moorage, and facilities, which depend in turn on vessel types, sizes, and numbers. A series of other issues that may require consideration include climate change impacts, permitting and approval requirements, sediment movement, currents, water quality and flushing, and navigability in the entrance channel. Economics, land transportation, and land-based issues are not normally considered within coastal engineering. In the case of ports or harbors that accommodate ships and larger vessels, many of the same considerations apply, but with an increased emphasis on a consideration of ship berths and moorings.

1.4 Evolution of Coastal Engineering and Future Trends

While coastal infrastructure had been built over many centuries, initially for coastal defense and harbor protection, it was not until the late 1800s that coastal infrastructure design was increasingly based on engineering principles, coinciding with the development of the engineering profession at that time. From the early 1900s onwards and certainly by the 1930s, key aspects of coastal engineering had become increasingly evident. These included an increased reliance on linkages to applied mathematics, fluid mechanics, oceanography, geology, and other disciplines; the application of scientific principles and scientific studies to areas such as coastal infrastructure design and beach protection; the introduction of laboratory modeling; an increasing number of papers published on coastal engineering topics; the formulation of early guidelines relating to coastal infrastructure and beach preservation; and the establishment of related organizations and research programs. For detailed information on the history and evolution of coastal engineering in various countries, the reader is referred to Kraus (1996).

Coastal engineering may be regarded as having evolved fully into a distinct discipline by the early 1950s. In 1950, the *First Conference on Coastal Engineering* was held in Long Beach, California with 35 invited papers. This conference series, which became known as the *International Conferences on Coastal Engineering*, is now held biennially in locations worldwide, with several hundred papers published in each set of proceedings.

Since the 1950s, there has been a continual broadening of the scope of coastal engineering from an initial focus on civil works projects so as to now include the development and application of advanced technologies over a range of areas such as dredging, beach nourishment, structural design, and port infrastructure and planning; the restoration, protection, and enhancement of coastal wetlands and other habitats, in collaboration with environmental engineers, environmental scientists, and biologists; and coastal management, with links to planning, geography, law, and other disciplines.

Associated with this broadening, a number of trends have emerged over the last few decades and are expected to evolve further into the future. These trends have arisen largely in response to emerging challenges, technological advances, and shifting societal values. A number of these are indicated below.

Computer modeling. Since the 1960s, the advent of computers and computer modeling has continually transformed coastal engineering practice. Today, computer modeling encompasses the use of spreadsheets to perform coastal engineering calculations, and the use of sophisticated computer models that can now describe highly complex and wide-ranging aspects of coastal engineering. This trend is expected to evolve further in the future. For example, AI (artificial intelligence) is already being relied upon in modeling coastal engineering phenomena.

Response to hazards and failures. Responses to natural disasters and failures, including devastation and damage arising from tsunamis, hurricanes, extreme storm surges, and extreme storms, along with lessons learned, have been the impetus to the establishment of research programs and to many research advances and design improvements, including the development of comprehensive design manuals that support engineering practice. As one of many examples, while breakwaters were being built in deeper water with larger artificial units in the 1960s and 1970s, extensive damage to a major breakwater at Port Sines, Portugal in 1978, which was associated with the strength, material properties, and shape of individual armor units, led to a major reconsideration of the design and deployment of very large armor units.

Probabilistic design. Another shift has placed a greater emphasis on approaches to accommodating uncertainties in the natural environment and in the design process. This has led to a more consistent approach to accounting for uncertainties in a project, an increased emphasis on probabilistic design, and the incorporation of risk assessment and risk management in projects.

Climate change. Of course, climate change and its impacts are now universally recognized, and approaches to accommodating the impacts of climate change have become an integral part of coastal engineering design. This is especially true with respect to the impacts of future sea level rise. Knowledge of climate change impacts is continually improving with respect to sea level rise, the intensity and behavior of hurricanes, extended ice-free seasons in Arctic regions, and wind and wave climates in different regions around the globe. Likewise, approaches to addressing these impacts, including adaptation measures along shorelines and probabilistic approaches to incorporating uncertainties, continue to evolve.

Range of applications. The range of applications of coastal engineering has been continually broadening, so that coastal engineers are increasingly engaged in areas such as the development of offshore renewable energy projects (involving also ocean engineering and other engineering disciplines), the restoration, protection, and enhancement of coastal wetlands and other habitats, and coastal management. Associated with this broadening are increased linkages with other disciplines such as geography, law, biology, business, and management, and increased interactions with planners, landscape architects, lawyers, biologists, developers, and government professionals. Three areas where such a broadening is occurring are highlighted below.

Nature-based shoreline protection. Associated with a greater recognition of the need to minimize environmental impacts, preserve habitats, and practice environmental stewardship, there has been an increased recognition of the need to incorporate coastal resilience into coastal engineering projects and to rely increasingly on "nature-based methods" of shoreline protection. Regulatory and permitting requirements increasingly require such considerations. While nature-based methods alone may be insufficient to provide adequate protection from shoreline erosion and flooding, the development of hybrid protection schemes that incorporate elements of traditional methods is becoming more common.

Coastal restoration. In a similar vein, there has been a major effort to support the recovery of degraded ecosystems and to support habitat protection and enhancement. This has led to a greater focus on the restoration of wetlands, salt marshes, and other coastal areas that support natural habitats. Engineers engaged in coastal restoration projects often collaborate with environmental biologists, landscape architects, and other professionals.

Coastal management and decision-making. Finally, coastal management, which refers to management activities relating to the coastal zone, is increasingly relied upon to assure an integrated and managed approach to potential interventions over broader areas of the coastal zone, with an emphasis on seeking sustainable solutions that are socially and environmentally responsible. Related to this, decision-making with respect to coastal development has become more inclusive and more reliant upon the engagement of user communities and all stakeholders. Accordingly, coastal engineering is becoming increasingly dependent on management, decision-making processes, and community and stakeholder interactions.

where FSBC denotes a free surface boundary condition. In the above, permanent form refers to the requirement that the wave train moves steadily without its shape changing.

The boundary value problem defined by the above set of governing equations is nonlinear and its solution is rather complicated. This is because, first, the two FSBC's apply on the initially unknown free surface location $z = \eta$, and, second, because the FSBC's contain other nonlinearities corresponding to products of variables or their derivatives.

2.3 Linear Wave Theory

2.3.1 Governing Equations

An additional assumption is made in order to simplify the complete boundary value problem so as to enable a relatively straightforward and robust solution to be developed. This assumption is that the wave height H is small in relation to the other length scales of the wave flow, L and d. That is, $H \ll L$, d. This simplification of the complete wave theory is referred to as linear wave theory, small amplitude wave theory, sinusoidal wave theory, or Airy wave theory.

There are two consequences of this additional assumption. First, the two free surface boundary conditions that apply on the initially unknown free surface location $z = \eta$ may instead be applied directly at the known SWL, $z = 0$. Second, the nonlinear terms in the two free surface boundary conditions, corresponding to products of variables or their derivatives, are an order of magnitude smaller than the remaining terms and are therefore neglected.

Applying these to the complete boundary value problem, the governing equations for linear wave theory may be developed as follows:

Laplace equation: $$\frac{\partial^2 \phi}{\partial x^2} + \frac{\partial^2 \phi}{\partial z^2} = 0 \qquad \text{within fluid region}$$

Seabed condition: $$\frac{\partial \phi}{\partial z} = 0 \qquad \text{at } z = -d$$

Kinematic FSBC: $$\frac{\partial \phi}{\partial z} = \frac{\partial \eta}{\partial t} \qquad \text{at } z = 0$$

Dynamic FSBC: $$\frac{\partial \phi}{\partial t} + g\eta = 0 \qquad \text{at } z = 0$$

Permanent form: $$\phi(x, z, t) = \phi(x - ct, z)$$

where again FSBC denotes a free surface boundary condition. The two free surface boundary conditions may be rewritten such that one equation excludes η so that this equation along with the remaining governing equations may be solved directly for ϕ, while a second equation provides an explicit expression for η in terms of ϕ. These are, respectively:

$$\frac{\partial^2 \phi}{\partial t^2} + g\frac{\partial \phi}{\partial z} = 0 \qquad \text{at } z = -d$$

$$\eta = -\frac{1}{g}\left(\frac{\partial \phi}{\partial t}\right)_{z=0}$$

2.3.2 Solution for Flow Field

The solution to the boundary value problem may readily be developed in order to obtain expressions for η and ϕ as follows:

$$\eta = \frac{H}{2}\cos(kx - \omega t)$$

$$\phi = \frac{\pi H}{kT}\frac{\cosh(ks)}{\sinh(kd)}\cos(kx - \omega t)$$

Corresponding expressions for the various flow parameters (u, w, …) may then be developed from the above expression for ϕ.

However, for a given depth d, the wave period T and wave length L are related, so that either one, but not both, needs to be specified in defining a wave train. Thus, the solution provides also the *linear dispersion relation* that relates L and T on the basis of linear wave theory, so that one can be obtained if the other is known. Equivalently, the dispersion relation relates ω and k, or c and k. The various results of linear wave theory are given in Table 2.1.

Table 2.1 Results of linear wave theory.

Variable	Equation
Free surface elevation	$\eta = \frac{H}{2}\cos(kx - \omega t)$
Velocity potential	$\phi = \frac{\pi H}{kT}\frac{\cosh(ks)}{\sinh(kd)}\sin(kx - \omega t)$
Dispersion relation	$\omega^2 = gk\tanh(kd)$ or
	$c^2 = \frac{g}{k}\tanh(kd)$
Horizontal displacement	$\xi = -\frac{H}{2}\frac{\cosh(ks)}{\sinh(kd)}\sin(kx - \omega t)$
Vertical displacement	$\zeta = \frac{H}{2}\frac{\sinh(ks)}{\sinh(kd)}\cos(kx - \omega t)$
Horizontal velocity	$u = \frac{\pi H}{T}\frac{\cosh(ks)}{\sinh(kd)}\cos(kx - \omega t)$
Vertical velocity	$w = \frac{\pi H}{T}\frac{\sinh(ks)}{\sinh(kd)}\sin(kx - \omega t)$
Horizontal acceleration	$\dot{u} = \frac{2\pi^2 H}{T^2}\frac{\cosh(ks)}{\sinh(kd)}\sin(kx - \omega t)$
Vertical acceleration	$\dot{w} = -\frac{2\pi^2 H}{T^2}\frac{\sinh(ks)}{\sinh(kd)}\cos(kx - \omega t)$
Pressure	$p = -\rho g z + \frac{\rho g H}{2}\frac{\cosh(ks)}{\cosh(kd)}\cos(kx - \omega t)$

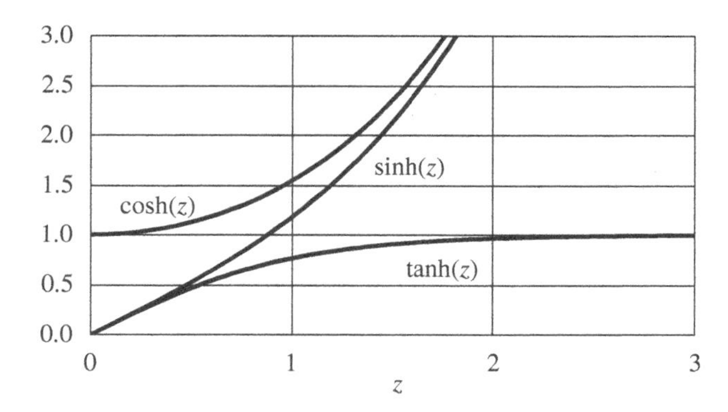

Figure 2.2 Hyperbolic functions.

Background on hyperbolic functions. The hyperbolic functions sinh, cosh, and tanh appear extensively in Table 2.1. As background, summary information on these functions is now provided. The relevant hyperbolic functions are defined in terms of an argument z as follows:

Hyperbolic sine: $$\sinh(z) = \frac{1}{2}\,[\exp(z) - \exp(-z)]$$

Hyperbolic cosine: $$\cosh(z) = \frac{1}{2}\,[\exp(z) + \exp(-z)]$$

Hyperbolic tangent: $$\tanh(z) = \frac{\sinh(z)}{\cosh(z)} = \frac{\exp(z) - \exp(-z)}{\exp(z) + \exp(-z)}$$

Each varies with the argument z as shown in Figure 2.2. Approximations for small and large values of the argument z are as follows:

For small z: $\sinh(z) \simeq \tanh(z) \simeq z$; $\cosh(z) \simeq 1$

For large z: $\sinh(z) \simeq \cosh(z) \simeq \frac{1}{2}\exp(z)$; $\tanh(z) \simeq 1$

2.3.3 Depth Parameter

The wave flow depends notably on the relative depth of the water, that is, the depth relative to the wave length, ranging from shallow water to deep water. Thus, a *depth parameter* kd ($= 2\pi d/L$), sometimes referred to as a *relative depth parameter*, appears in many of the formulae in Table 2.1. It is of interest to consider cases where the depth parameter is large, corresponding to deep-water waves, and is small, corresponding to shallow-water waves. Suitable approximations may be made for each of these ranges, leading to simplifications to the expressions given in Table 2.1. Between these two limiting cases, the waves are termed *intermediate-depth waves*, with the complete expressions in Table 2.1 then used. The deep-water and shallow-water approximations are now considered.

Deep-water waves. Deep-water waves correspond to conditions described by $kd > \pi$ (i.e. $d/L > 1/2$). Large argument approximations to the hyperbolic functions then lead to simplified expressions in which

Table 2.2 Results of linear wave theory – deep water.

Variable	Equation
Free surface elevation	$\eta = \frac{H}{2}\cos(kx - \omega t)$
Velocity potential	$\phi = \frac{\pi H}{kT}\exp(kz)\sin(kx - \omega t)$
Dispersion relation	$\omega^2 = gk$ or $c^2 = \frac{g}{k}$
Horizontal displacement	$\xi = -\frac{H}{2}\exp(kz)\sin(kx - \omega t)$
Vertical displacement	$\zeta = \frac{H}{2}\exp(kz)\cos(kx - \omega t)$
Horizontal velocity	$u = \frac{\pi H}{T}\exp(kz)\cos(kx - \omega t)$
Vertical velocity	$w = \frac{\pi H}{T}\exp(kz)\sin(kx - \omega t)$
Horizontal acceleration	$\dot{u} = \frac{2\pi^2 H}{T^2}\exp(kz)\sin(kx - \omega t)$
Vertical acceleration	$\dot{w} = -\frac{2\pi^2 H}{T^2}\exp(kz)\cos(kx - \omega t)$
Pressure	$p = -\rho gz + \frac{\rho gH}{2}\exp(kz)\cos(kx - \omega t)$

the depth d is absent as expected. The subscript "o" is used to denote deep-water wave conditions. The dispersion relation may be simplified to the following alternatives:

$$L_o = \frac{gT^2}{2\pi}; \qquad c_o = \frac{gT}{2\pi}; \qquad c_o = \sqrt{\frac{gL}{2\pi}}$$

Thus, useful conversion formulae for L_o in terms of T include $L_o\,(\mathrm{m}) = 1.56\,T^2$ and $L_o\,(\mathrm{ft}) = 5.12T^2$. Based on the above approximations, the various results of linear wave theory for deep-water waves are given in Table 2.2.

Shallow-water waves. Shallow-water waves correspond to conditions described by $kd < \pi/10$ (i.e. $d/L < 1/20$). Small argument approximations to the hyperbolic functions can be introduced and the dispersion relation may be simplified to:

$$c = \sqrt{gd}$$

However, beyond this simplification for c, it is convenient to continue to rely on the complete expressions for the various parameters given in Table 2.1.

2.3.4 Description of Results

The flow corresponding to Table 2.1 is now described in terms of the water particle orbits and the variations of the amplitudes of u and w with elevation. Thus, Figure 2.3 provides sketches of these for shallow-water waves, intermediate-depth waves, and deep-water waves.

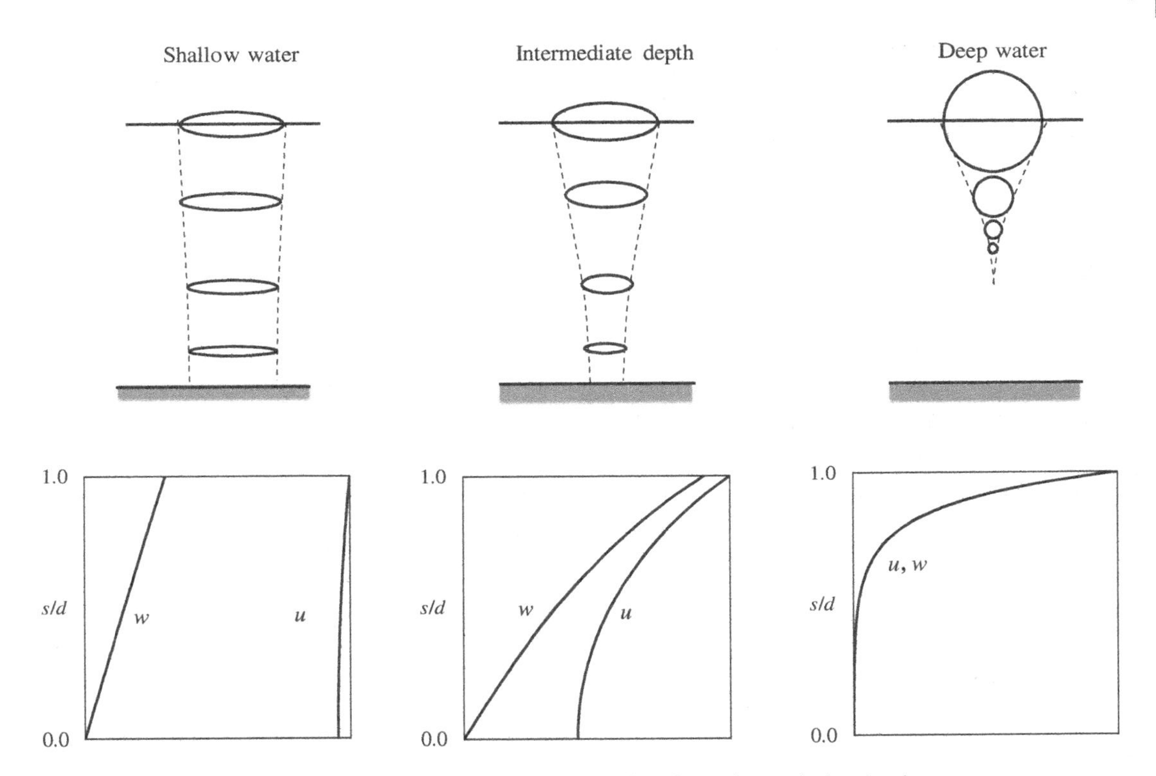

Figure 2.3 Water particle orbits and velocity amplitude profiles for various relative depths.

The figure indicates how the water particles have elliptic orbits, with diameters that decay with reducing z or s. These range from flatter elliptic orbits for shallow-water waves, which are influenced by the seabed and which decay only gradually with depth, to circular orbits for deep-water waves, which are not influenced by the seabed and which decay with depth so as to vanish at close to half a wave length below the water surface.

2.3.5 Linear Dispersion Relation

The linear dispersion relation relates L and T (or ω and k) as indicated in Table 2.1. If L (or k) is known, then T (or ω) may be obtained directly from:

$$\omega^2 = gk \tanh(kd)$$

However, if T (or ω) is known, which is more often the case, then L (or k) needs to be obtained by an iterative or other approximate method. This is a fundamental requirement in most analyses of wave conditions. Three methods that may be adopted are outlined below.

Iterative method. One approach is to recast the dispersion relation into the form:

$$kd \tanh(kd) = \frac{4\pi^2 d}{gT^2}$$

Table 2.3 Corresponding values of kd and d/gT^2 for kd estimation.

d/gT^2	kd	d/gT^2	kd	d/gT^2	kd
Shallow-water limit:		0.018	0.957	0.060	2.407
$kd = 2\pi\sqrt{d/gT^2}$		0.020	1.024	0.065	2.595
0.001	0.200	0.022	1.090	0.070	2.785
0.002	0.285	0.024	1.156	0.075	2.976
0.003	0.351	0.026	1.222	0.080	3.169
0.004	0.408	0.028	1.288	0.085	3.364
0.006	0.507	0.030	1.354	0.090	3.559
0.008	0.593	0.035	1.520	0.095	3.755
0.010	0.673	0.040	1.690	0.100	3.951
0.012	0.748	0.045	1.864	0.105	4.147
0.014	0.819	0.050	2.042	Deep-water limit:	
0.016	0.889	0.055	2.223	$kd = 4\pi^2\ (d/gT^2)$	

This can readily be solved iteratively for kd using:

$$X_{n+1} = X_n - \frac{X_n \tanh(X_n) - A}{\tanh(X_n) + X_n \operatorname{sech}^2(X_n)}$$

with $X = kd$, $A = 4\pi^2(d/gT^2)$, and X_o taken as A or $\sqrt{A}$.

Look-up table. A simple approach is to rely on interpolation using Table 2.3, which shows corresponding kd and d/gT^2 values. Exact formulae for kd for the shallow-water and deep-water cases are available and are included in the table.

Regression fit. Thirdly, a regression fit to various ranges of the exact equation has provided the following equation for kd in terms of d/gT^2:

$$kd = \begin{cases} X & \text{for } X < 0.15 \\ 0.3228\,X^3 - 0.1919\,X^2 + 1.0794\,X - 0.0102 & \text{for } 0.15 < X < 1.09 \\ 0.0620\,X^3 + 0.9891\,X^2 - 0.5891\,X + 0.7451 & \text{for } 1.09 < X < 1.77 \\ -0.1147\,X^3 + 1.7502\,X^2 - 1.6431\,X + 1.2055 & \text{for } 1.77 < X < 2.24 \\ X^2 & \text{for } X > 2.24 \end{cases}$$

where $X = 2\pi\sqrt{d/gT^2}$.

Reference Solution A1 in Appendix A provides a spreadsheet solution to the linear dispersion relationship based on the iterative and regression fit methods. (The look-up table method is the simplest one to use without a spreadsheet.) This spreadsheet solution also includes the case of a coexisting current, to be considered in Section 2.5.

Example 2.1 ***Application of Linear Wave Theory***

A wave train has a wave height $H = 1.6$ m and wave period $T = 3.7$ s at a location where the still water depth $d = 7$ m. On the basis of linear wave theory, calculate the wave length, the wave speed, the maximum horizontal velocity at mid-depth, and the orbital diameter of water particle motions at the seabed.

Solution

Specified parameters

$g = 9.80665\ \text{m/s}^2$
$H = 1.6\ \text{m}$
$T = 3.7\ \text{s}$
$d = 7.0\ \text{m}$

Solve for kd (three methods shown)

$d/gT^2 = 0.0521$
Method 1 – iteration method: $kd = 2.1188$
Method 2 – look-up table: $kd = 2.1190$
Method 3 – regression fit: $kd = 2.1193$

Wave length and speed

$k = kd/d = 0.3027/\text{m}$
$L = 2\pi/k = 20.8\ \text{m}$
$c = L/T = 5.6\ \text{m/s}$

Maximum horizontal velocity at mid-depth

A formula for the required maximum velocity u_m may be developed from the formula for the horizontal velocity u given in Table 2.1 by taking $\cos(kx - \omega t) = 1$ for the amplitude or maximum value and $s = d/2$ for the mid-depth value. Thus:

$$u_m = \left(\frac{\pi H}{T}\right)\left[\frac{\cosh(kd/2)}{\sinh(kd)}\right] = 0.5\ \text{m/s}$$

Orbital diameter of water particle motions at the seabed

A formula for the required diameter d_o may be developed from the formula for the horizontal displacement ξ given in Table 2.1 by taking $\cos(kx - \omega t) = 1$ for the amplitude or maximum value and $s = 0$ for the value at the seabed, and then doubling this to convert from amplitude to diameter. Thus:

$$d_o = 2\left(\frac{H}{2}\right)\left[\frac{\cosh(0)}{\sinh(kd)}\right] = 0.4\ \text{m}$$

2.4 Wave Energy and Momentum

A description of the energy and momentum characteristics of a wave train is also needed, especially with respect to the propagation of groups of waves, the development of wave transformation relationships to be considered in Chapter 3, and the behavior of waves under various circumstances. Two energy-related parameters are defined as follows:

Wave energy density, E: average energy per unit horizontal area due to waves (in units of J/m^2 or N/m)
Wave energy flux, P: rate of energy propagation per unit width in the wave direction (in units of J/ms or N/s)

An expression for E may be derived by considering the kinetic and potential energy contained in an element of height dz and horizontal area $dx\,dy$, integrating this from the seabed to the instantaneous water surface, taking the time average of the result, and then subtracting the corresponding quantity in the absence of waves. That is, an expression for E may be developed from:

$$E = \overline{\int_{-d}^{\eta} \left[\frac{1}{2}\rho(u^2 + w^2) + \rho g z\right] dz} - \overline{\int_{-d}^{0} \rho g z \, dz}$$

where an overbar denotes a time average.

In a similar manner, an expression for P may be derived by considering the instantaneous rate at which work is done on, and kinetic and potential energy is transferred across, a vertical element of height dz, integrating this from the seabed to the instantaneous water surface, and then taking the time average of the result. That is, an expression for P may be developed from:

$$P = \overline{\int_{-d}^{\eta} \left[p + \frac{1}{2}\rho(u^2 + w^2) + \rho g z\right] u \, dz}$$

When the results of linear wave theory are substituted into the above expressions and the integrations are carried out, the final expressions for E and P simplify to the following:

$$E = \frac{1}{8}\rho g H^2$$

$$P = \frac{1}{16}\rho g H^2 c \left[1 + \frac{2kd}{\sinh(2kd)}\right]$$

From these, the *group velocity* c_G is defined as $c_G = P/E$ and thus is given by the expression:

$$c_G = \frac{P}{E} = \frac{c}{2}\left[1 + \frac{2kd}{\sinh(2kd)}\right]$$

The group velocity corresponds to the speed of wave energy propagation that is equal to the velocity of propagation of a group of waves, as distinct from the velocity of the individual waves within the group. For deep-water waves, this implies that individual waves travel at the wave speed c, whereas the group itself travels at the group velocity $c_G = c/2$. This corresponds to individual waves moving towards the front of the group and then attenuating, while new waves form at the rear of the group and advance relative to the group.

This has the somewhat paradoxical implication that, for a short group of deep-water waves, the spatial variation of η at a particular instant t_1 and the time variation of η at a particular location x_1 will have the

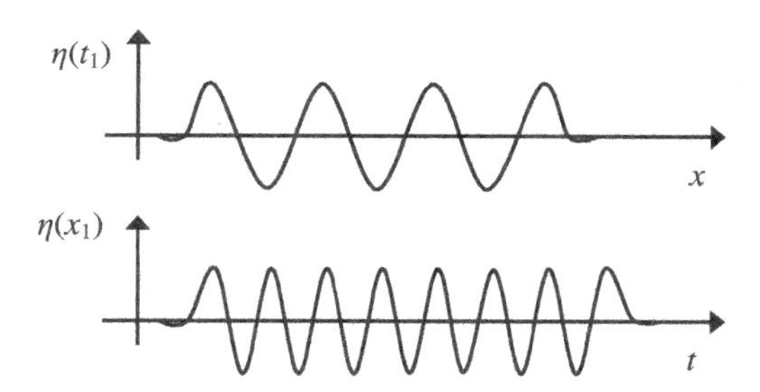

Figure 2.4 Spatial and time variations of a wave group.

profiles as indicated in Figure 2.4, with the former exhibiting half as many waves as the latter. This can be observed in a laboratory setting and can readily be explained.

In a similar manner to the consideration of wave energy outlined above, parameters describing the momentum of a wave train may also be developed. Specifically, the *radiation stress* S corresponds to the excess flux of momentum due to the presence of the waves. However, since momentum is a vector, the radiation stress is described by a tensor with four elements reflecting directional aspects of the transfer of momentum. Thus, the elements S_{xx} and S_{xy} represent the flow of x-ward momentum transported in the x and y directions, respectively, while S_{yx} and S_{yy} represent the flow of y-ward momentum transported in the x and y directions, respectively.

When the results of linear wave theory are substituted into the relevant expressions for the radiation stresses and the integrations are carried out, the final expressions for S_{xx} and S_{yy} simplify to the following:

$$S_{xx} = \left[\frac{1}{2} + \frac{2kd}{\sinh(2kd)}\right] E$$

$$S_{yy} = \left[\frac{kd}{\sinh(2kd)}\right] E$$

while $S_{xy} = S_{yx} = 0$. It is noted that, even though the y-ward velocity v is continually zero, S_{yy} is nonzero because of a nonzero pressure term in the integral expression for radiation stress.

The radiation stress has applications to various wave phenomena that are associated with wave amplitude changes as a wave train propagates. These include wave interactions with a non-uniform current, surf beats, which are long-period oscillations associated with a wave group reaching the shore, wave setup, which is a change in mean water level as waves break and reach the shore, and longshore currents, which are generated parallel to the shore as waves approach the shore obliquely.

2.5 Waves with a Current

Consideration is now given to a wave train propagating in the x direction in the presence of an underlying, uniform current U that flows in a direction α relative to the wave direction. This is distinct from the treatment of wave–current interactions, not considered here, whereby waves are modified when they initially encounter a current field.

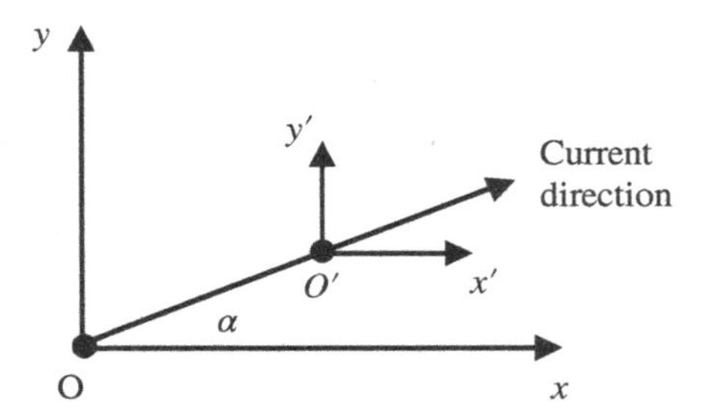

Figure 2.5 Coordinate systems for waves with a current.

2.5.1 Fixed and Moving Reference Frames

In analyzing coexisting waves and currents, reference is made to the fixed reference frame (O, x, y) in the horizontal plane, as well as a reference frame that moves with the current, denoted (O', x', y'), as indicated in Figure 2.5.

The moving reference frame moves at a speed $U\cos\alpha$ in the x direction, so that the wave speed relative to the fixed reference frame, denoted c_{f}, is given in terms of the wave speed c relative to the current (i.e. the moving reference frame) by:

$$c_{\mathrm{f}} = c + U\ \cos\alpha$$

The wave height H, wave length L, and wave number k do not depend on the choice reference frame, whereas the wave period T and wave angular frequency ω do, so that the subscript f may be applied to the latter parameters in the same way. In particular, by substituting $c_{\mathrm{f}} = \omega_{\mathrm{f}}/k$ and $c = \omega/k$ in the above, ω_f and ω are related by:

$$\omega_{\mathrm{f}} = \omega + kU\ \cos\alpha$$

In fact, the wave speed and the wave flow relative to the current are identical to those for the wave train if there was no current present. Therefore, the usual dispersion relation applies between c and L and between ω and k (i.e. relative to the moving reference frame), but not between c_{f} and L or between ω_{f} and k (i.e. relative to the fixed reference frame).

2.5.2 Solution for Flow Field

Based on the above, the various formulae for the case of no current given in Table 2.1 may be suitably extended so as to describe the flow relative to a fixed reference frame when a current is present. Overall, the flow is now described by the following:

$$\eta = \frac{H}{2}\cos(kx - \omega_{\mathrm{f}}t)$$

$$\phi = \frac{\pi H}{kT}\frac{\cosh(ks)}{\sinh(kd)}\sin(kx - \omega_{\mathrm{f}}t) + Ux\ \cos\alpha + Uy\ \sin\alpha$$

$$u = \frac{\pi H}{T}\frac{\cosh(ks)}{\sinh(kd)}\cos(kx - \omega_{\mathrm{f}}t) + U\ \cos\alpha$$

$$v = U\ \sin\alpha$$

$$w = \frac{\pi H}{T}\frac{\sinh(ks)}{\sinh(kd)}\ \sin(kx - \omega_f t)$$

It is seen that, in the presence of the current, the velocity u in the x direction has a steady current component plus an oscillatory wave component (with angular frequency ω_f, not ω), the velocity v in the y direction has a steady current component only, and the velocity w in the z direction has an oscillatory wave component only (again with angular frequency ω_f, not ω).

2.5.3 Dispersion Relation

When T (relative to the current) is known, the usual dispersion relation between ω and k may be used to obtain k. The above equation for ω_f may then be applied so that all the details of the flow may be determined. However, in the more common case when T_f (relative to a fixed reference frame) is known, a different dispersion relation linking ω_f and k also involves the current magnitude and is less straightforward to apply. In particular, kd may then be obtained by a modified dispersion relation that may be written as:

$$kd\ \tanh(kd) = \frac{\omega_f^2 d}{g}\left[1 - \left(\frac{U\ \cos\alpha}{\omega_f d}\right)kd\right]^2$$

This can be solved iteratively for kd using:

$$X_{n+1} = X_n - \frac{X_n\ \tanh(X_n) - A + BX_n + CX_n^2}{\tanh(X_n) + X_n \mathrm{sech}^2(X_n) + B - 2CX_n}$$

with $X = kd$, $A = \omega_f^2 d/g$, $B = 2A(U\ \cos\alpha/\omega_f d)$, $C = B^2/4A$, and X_o taken as A or $\sqrt{A}$.

Reference Solution A1 in Appendix A includes a spreadsheet solution to the dispersion relation with a current based on the above formulation.

Note that in deep water there is no need for an iterative solution. Instead, when both U and ω_c are known, k may be obtained as a solution to the following quadratic equation:

$$(U^2\cos^2\alpha)k^2 - (2\omega_f U\ \cos\alpha + g)k + \omega_f^2 = 0$$

Example 2.2 ***Waves on a Uniform Current***
A wave-current field corresponds to $H = 2$ m, $T_f = 6$ s (relative to a fixed frame), $d = 10$ m, $U = 2$ m/s, and $\alpha = 150°$ (i.e. the waves propagate obliquely against the current). On the basis of linear wave theory, calculate the wave length, the wave speed relative to a fixed reference frame, the maximum water particle velocity at the seabed, and the direction of this maximum particle velocity relative to the wave direction.

Solution

Specified parameters

$g = 9.80665$ m/s^2
$H = 2.0$ m

$T_f = 6.0$ s
$d = 10.0$ m
$U = 2.0$ m/s
$\alpha = 150°$

Solve for *kd*

The Reference Solution A1 spreadsheet provides kd in terms of d, g, T_f, U, and α. This yields:

$kd = 2.08$

Wave length and speed

$k = kd/d = 0.208/\text{m}$
$L = 2\pi/k = 30.2$ m
$c_f = L/T_f = 5.0$ m/s

The wave period T (relative to the current) is needed in the velocity formula and is obtained via ω as follows:

$$\omega = \sqrt{gk\ \tanh(kd)} = 1.41 \text{ rad/s}$$

$$T = 2\pi/\omega = 4.5 \text{ s}$$

Maximum velocity

The maximum velocity is obtained as $V = \sqrt{u_m^2 + v_m^2}$, where u_m and v_m are the maximum velocities in the x and y directions, respectively. (Since v_m is steady, there is no need to consider phases in this formula.) Based on the formulae for u and v, taking $\cos(kx - \omega_f t) = 1$ for the maximum value, and taking $s = 0$ for the seabed, u_m, v_m, and V are obtained as:

$$u_m = -\left(\frac{\pi H}{T}\right)\left[\frac{1}{\sinh(kd)}\right] + U\ \cos\alpha = -2.1 \text{ m/s}$$ (i.e. u_m is a maximum in the negative x direction)

$$v_m = U \sin\alpha = 1.0 \text{ m/s}$$

$$V = \sqrt{u_m^2 + v_m^2} = 2.3 \text{ m/s}$$

The direction of V relative to the wave direction, α_V, is given as:

$$\alpha_V = \text{acos}(u_m/V) = 154°$$

■➔

2.6 Extensions to Linear Wave Theory

A number of extensions and alternate formulations of linear wave theory are now outlined.

2.6.1 Waves Propagating at an Angle to the *x* axis

First, while the propagation of waves in the x direction has already been considered, it is relatively straightforward to consider waves that propagate in some other direction α relative to the x axis. With

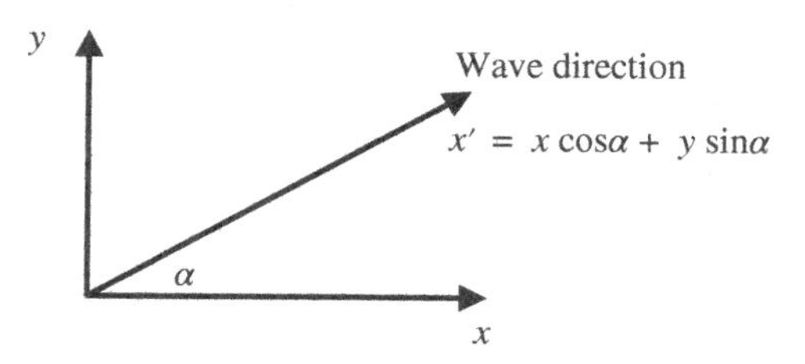

Figure 2.6 Coordinate system for oblique wave propagation.

reference to Figure 2.6, wave theory applies in the usual way with respect to the direction x' instead of x.

By expressing x' in terms of x and y, as shown in the figure, the results of wave theory may be expressed in terms of x and y in place of x' alone. In particular:

$$\eta = \frac{H}{2}\cos(kx\cos\alpha + ky\sin\alpha - \omega t)$$

$$\phi = \frac{\pi H}{kT}\frac{\cosh(ks)}{\sinh(kd)}\sin(kx\cos\alpha + ky\sin\alpha - \omega t)$$

Expressions for other flow parameters may be developed accordingly.

2.6.2 Reference Frame Moving with the Waves

In some cases, it is useful to consider the development of wave theories in the context of the flow relative to a reference frame (O', x', z) that moves at a speed c with the waves, rather than one that is fixed, (O, x, z). This distinction is illustrated in Figure 2.7.

x' is given as $x' = x - \text{ct}$, such that a wave crest continually lies at $x' = 0$. Relative to this reference frame (O', x', z), the flow is steady so that the various variables do not then vary with time t. (A steady flow is one in which flow variables may vary with location, but not with time.) Given that the reference frame moves at a speed c and the flow is steady, the horizontal velocity u' relative to the moving reference frame is given in terms of the horizontal velocity u relative the fixed reference frame as $u'(x', z) = u(x, z, t) - c$. Similarly, the corresponding velocity potential is given as $\phi'(x', z) = \phi(x, z, t) - \text{cx}$.

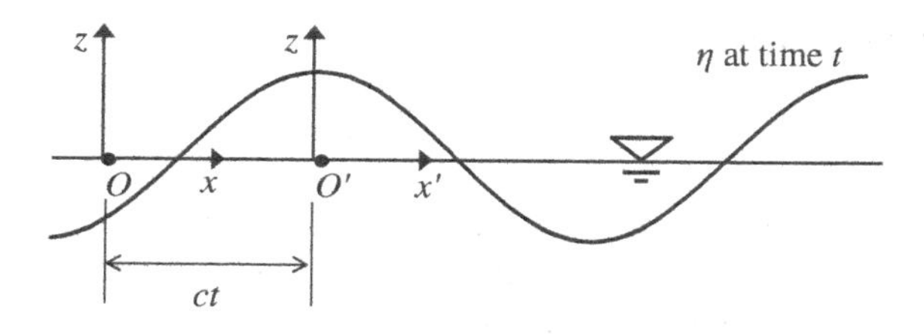

Figure 2.7 Comparison of reference frames.

Relative to the reference frame (O', x', z), time-dependent terms may be omitted so that the governing equations for the complete boundary value problem are as follows:

Laplace equation: $\frac{\partial^2 \phi'}{\partial x'^2} + \frac{\partial^2 \phi'}{\partial z^2} = 0$ within the fluid region, at $z = -d$

Seabed condition: $\frac{\partial \phi'}{\partial z} = 0$

Kinematic FSBC: $\frac{\partial \phi'}{\partial z} = \frac{\partial \phi'}{\partial x'} \frac{\partial \eta}{\partial x'}$ at $z = \eta$

Dynamic FSBC: $\frac{1}{2}(u'^2 + w^2) + g\eta = 0$ at $z = \eta$

In the case of linear wave theory, the solution may be developed as follows:

$$\eta = \frac{H}{2} \cos(kx')$$

$$\phi' = \frac{\pi H}{kT} \frac{\cosh(ks)}{\sinh(kd)} \sin(kx') - cx'$$

$$u' = \frac{\pi H}{T} \frac{\cosh(ks)}{\sinh(kd)} \cos(kx') - c$$

2.6.3 Stream Function Representation

In the development of wave theory described in Section 2.2, the velocity components u and w were combined into a single variable ϕ on the basis of the irrotationality condition, with the continuity equation then used to develop the Laplace equation for ϕ.

In a complementary approach, u and w may be combined into a different variable, termed the stream function ψ. This time, ψ is defined on the basis of the continuity condition, so that the irrotationality condition is instead used to develop the Laplace equation for ψ. Thus, the stream function is defined by:

$$u = \frac{\partial \psi}{\partial z}; w = -\frac{\partial \psi}{\partial x} \quad \left[\text{c.f.} : u = \frac{\partial \phi}{\partial x}; w = \frac{\partial \phi}{\partial z}\right]$$

Linear wave theory may then be developed in terms of ψ rather than ϕ, with the resulting solution for ψ given as:

$$\psi = \frac{\pi H}{kT} \frac{\sinh(ks)}{\sinh(kd)} \cos(kx - \omega t)$$

For certain situations, it may be advantageous to work with ψ rather than ϕ. However, it is noted that, unlike ϕ, ψ cannot be extended to general three-dimensional flows.

2.6.4 Complex Representation

For some extensions of linear wave theory, it is useful to recast the various equations using complex variables, with the real part of any complex quantity corresponding to a physical realization of that quantity. This is especially useful when considering combinations of wave components with different or changing wave phases.

As a starting point, it is noted that Euler's formula is:

$$\exp(ix) = \cos(x) + i \sin(x)$$

where $i = \sqrt{-1}$ and x is real. Based on this relationship, a sinusoidal signal $a \cos(\omega t - \alpha)$ with amplitude a and phase α may be represented in complex form as $A \exp(-i\omega t)$, where A is the complex amplitude of the signal. Using this representation, it may be shown that $a = |A|$, the modulus of A, and $\alpha = \text{Arg}(A)$, the argument of A.

Using this complex notation, the expressions for η and ϕ given by linear wave theory are:

$$\eta = \frac{H}{2} \exp[i(kx - \omega t)]$$

$$\phi = \frac{\pi H}{kT} \frac{\cosh(ks)}{\sinh(kd)} \exp[i(kx - \omega t)]$$

Examples of the use of complex notation are provided in Sections 6.5.2 and 8.5.2.

2.7 Nonlinear Wave Theories

Linear wave theory has been developed on the basis of a small wave height assumption, $H \ll L, d$. It may be desirable to avoid this assumption in order to develop a more accurate description of waves with larger heights. In particular, waves with large heights in shallower water have steeper crests and flatter troughs relative to those of a sinusoidal profile, as illustrated in Figure 2.8. Because of this, a sinusoidal representation of the wave profile and of flow variations may be insufficiently accurate. This has led to the development of nonlinear wave theories, such that the linear approximations to the two free surface boundary conditions are no longer relied upon.

Three families of nonlinear wave theories have been developed as follows:

- Stokes wave theories
- cnoidal wave theories, including the limiting case of solitary wave theories
- numerical wave theories

Summaries of these theories are given below. The Stokes and cnoidal wave theories are no longer used in coastal engineering practice, since numerical wave theories that describe the flow for steep waves over a full range of relative depths are now available.

2.7.1 Stokes Wave Theories

The Stokes wave theories, originally developed by Stokes in 1847, provide an analytical solution to the governing equations by representing ϕ and η as perturbation series. Each is expressed as a power series

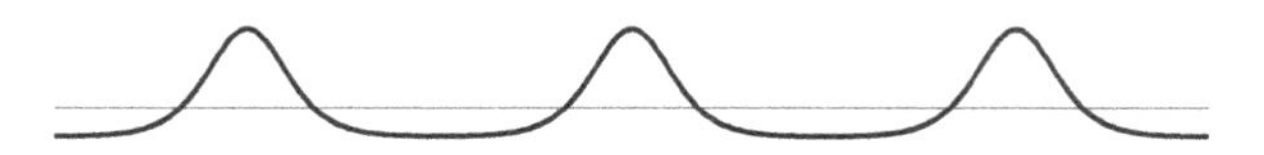

Figure 2.8 Profile of steep waves in shallow water.

in a small perturbation parameter ε, so that each successive term is an order of magnitude smaller than the preceding one:

$$\eta = \varepsilon\,\eta_1 + \varepsilon^2\,\eta_2 + \cdots$$

$$\phi = \varepsilon\,\phi_1 + \varepsilon^2\,\phi_2 + \cdots$$

Taylor series expansions of the two nonlinear free surface boundary conditions about the SWL are then applied in order to determine conditions at the SWL. Substituting the perturbation series into the various governing equations, and then collecting terms to order ε, ε^2, ε^3, ... in turn, a separate set of governing equations corresponding to each order of magnitude may be established. The solution to any particular order relies on the solutions for lower orders. The equations to order ε are identical to the linear theory problem already treated. The terms to order ε^2 provide Stokes second-order wave theory, and so on.

Second-order theory gives, for example:

$$\eta = \frac{H}{2}\cos(kx - \omega t) + \frac{H}{8}\left(\frac{\pi H}{L}\right)\left\{\frac{\cosh(kd)[2 + \cosh(2kd)]}{\sinh^3(kd)}\right\}\cos[2(kx - \omega t)]$$

ω and k continue to be related by the linear dispersion relation.

Second-order terms should be an order of magnitude smaller than the first-order terms. However, they may become unduly large in shallow water, so that Stokes second-order theory is then considered to fail. Stokes fifth-order wave theories were widely used in ocean engineering practice but have now been replaced by highly accurate numerical wave theories.

2.7.2 Cnoidal Wave Theories

As an alternative analytical solution suitable for shallow-water waves, $d \ll L$, cnoidal wave theories may be developed by initially "stretching" the vertical length scale relative to the horizontal length scale prior to the application of a perturbation series to represent the flow and of the Taylor series expansions to treat the free surface boundary conditions. The terminology "cnoidal" wave theory arises because the solution then involves the Jacobian elliptic function cn(). Figure 2.9 illustrates the wave profile based on cnoidal wave theory and shows the familiar long flat troughs and narrow crests of real waves in shallow water. The figure also includes its limiting cases of a solitary wave profile as $L/d \to \infty$ and a linear wave profile as $H/d \to 0$.

Cnoidal wave theories are not used in coastal engineering practice. However, the limiting case of a solitary wave is sometimes considered and this is now summarized.

2.7.3 Solitary Wave Theories

The limiting case of a solitary wave, corresponding to $L/d \to \infty$, may be developed by applying a perturbation series now in terms of the parameter H/d. The lowest order theory provides details of the flow as follows:

$$\frac{\eta}{d} = \frac{u}{\sqrt{gd}} = \frac{H}{d}\mathrm{sech}^2(\theta)$$

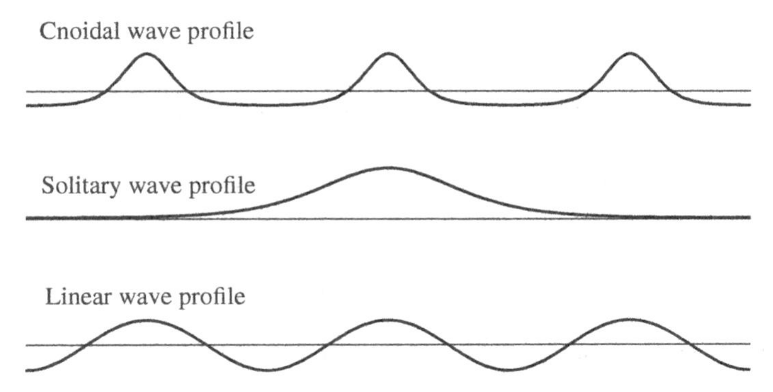

Figure 2.9 Comparison of shallow wave profiles.

$$\frac{\dot{u}}{g} = \frac{H}{d}\sqrt{3H/d}\,\text{sech}^2(\theta)\ \tanh(\theta)$$

$$c = \sqrt{gd}$$

with $\theta = k'(x - ct)$, $\text{sech}(\theta) = 1/\cosh(\theta)$, and $k' = \sqrt{3H/d}/2d$. Although higher order versions of solitary wave theory are available, they are not used in engineering practice.

2.7.4 Numerical Wave Theories

As mentioned, highly accurate numerical wave theories over all wave conditions are now available. Several variants of these have been developed, starting in the 1960s. These have usually relied on a stream function representation of the flow; hence, the earlier versions have been referred to as stream function wave theories, while subsequent versions are more commonly referred to as numerical wave theories. The following summary is based on the version described by Fenton (1988). In this approach, the flow is described in terms of the stream function ψ that was defined in Section 2.6.3 and is developed by solving the governing equations based on a reference frame moving with the waves. These equations are equivalent to those given in Section 2.6.2 with respect to the velocity potential. The stream function ψ is expressed as:

$$\psi(x', z) = -cs + \sqrt{\frac{g}{k^3}}\sum_{j=1}^{N} B_j \frac{\sinh(jks)}{\cosh(jkd)}\cos(jkx')$$

where $x' = x - ct$ is distance in the direction of wave propagation measured from an origin that moves with the waves, N is the number of terms of a Fourier approximation of the wave form, and B_j are unknown coefficients to be determined. The above equations satisfy the Laplace equation within the fluid and the seabed condition. To solve the problem numerically, the two free surface boundary conditions are satisfied at $N + 1$ discrete points extending over half a wave length from a crest to a trough, so as to provide the coefficients B_j. Finally, an additional assumption relating to the absence of a current is invoked in order to develop the dispersion relation between L and T. Once the coefficients B_j are obtained for specified values of H, L, and d, or H, T, and d, the various flow parameters may be determined.

Problems

2.1 A wave train with wave height $H = 8$ m and wave period $T = 11$ s propagates in water of uniform depth $d = 20$ m. On the basis of linear wave theory, determine the wave length, the wave speed, the maximum horizontal velocity at mid-depth, and the maximum horizontal acceleration at the SWL. Plot the distributions over depth (from the seabed to the SWL) of hydrostatic pressure, dynamic pressure amplitude, and the minimum and maximum total pressure. What are their values at the SWL and at the seabed? Assume that $\rho = 1025$ kg/m^3. *[Ignore pressure discontinuities at the SWL.]*

2.2 For deep-water waves, what is the particle orbital diameter at a distance $L/4$ below the SWL, expressed as a fraction of that at the SWL?

2.3 An earthquake in the Aleutian Islands (approximately 170°W) generates a tsunami (that is a shallow-water wave train). Use the Internet and Google Earth to estimate average depths and distances, and thereby determine the average speed and time of travel to Hawaii and to the west coast of Vancouver Island.

2.4 Swell (long, low-frequency waves) from a distant storm arrives at Los Angeles from the southwest. When the swell first arrives, the wave period is 20 s, and after 48 hr the wave period has changed to 15 s. Estimate when the swell was initially generated (in days before its initial arrival at Los Angeles) and the distance of the storm from Los Angeles. Assume that the waves generated by the storm, regardless of wave period, all start at the same location and time and that the swell travels at the group velocity for deep-water waves. Use Google Earth to determine the approximate location (longitude and latitude) of the storm.

2.5 Deep-water waves with a height $H = 0.5$ m and wave length $L = 20$ m propagate downriver towards the west in the same direction as a 1.2 m/s river current. What is the wave speed relative to an observer in a boat that is travelling at 2.0 m/s towards the northwest? What is the maximum water particle velocity at the SWL relative to a fixed reference frame?

2.6 ■ The wave height H of a wave train arising from the superposition of two wave trains propagating in the x direction with different heights H_1 and H_2, the same wave period, and a phase difference ε between them is required. By considering the expression for η using the complex representation given in Section 2.6.4, derive an expression for H in terms of H_1, H_2 and ε, written as the modulus of a complex expression. By using functions of complex numbers on EXCEL (e.g. IMABS, IMEXP), determine this wave height when $H_1 = 0.8$ m, $H_2 = 1.2$ m and $\varepsilon = 30°$. As an optional extension, develop a corresponding expression for H that does not rely on a complex representation and use this to obtain the corresponding result for H.

2.7 ■ By applying solitary wave theory to a wave with $H = 1.2$ m and $d = 5$ m, estimate the maximum fluid velocity, the maximum fluid acceleration, and the instants (in seconds) at which these occur after a wave crest crosses a given location.

3

Wave Transformations

This chapter considers wave transformations due to changes in water depth and the presence of barriers or obstacles in the flow, including wave shoaling, refraction, diffraction, reflection, attenuation, and breaking. Wave runup at a shoreline is also described. Many of the transformations described here are based on linear wave theory. Transformations based on nonlinear wave theories are used in the development of sophisticated numerical models and are summarized in Section 3.11.

3.1 Wave Shoaling

Wave shoaling refers to changes in a wave train associated with changes in water depth as the waves approach a shoreline normally (i.e. with wave crests that are parallel to the seabed contours). The fundamental situation is sketched in Figure 3.1. With reference to the figure, the intent is to develop a set of shoaling relations that relate conditions at some intermediate depth, described by H, L, c, and T, in terms of the corresponding parameters in deep water, denoted by the subscript o.

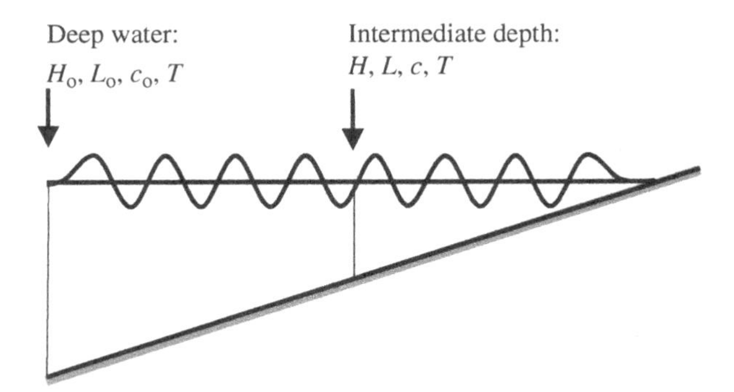

Figure 3.1 Definition sketch of wave shoaling.

An Introduction to Coastal Engineering, First Edition. Michael Isaacson.

Companion website: www.wiley.com/go/coastalengineering

3.1.1 Assumptions

The shoaling relations are developed on the basis of the following assumptions:

- ***Normal direction***. The waves approach a shoreline with straight parallel contours normally, so that the corresponding flow is two-dimensional in the (x, z) plane.
- ***Constant wave period***. As the waves approach the shoreline, the wave period is constant. That is, over a given duration, an equal number of waves cross a fixed point in deep water as those that cross another fixed point in shallower water.
- ***Constant wave energy flux***. As the waves approach the shoreline, the wave energy flux is constant. That is, it is assumed that there is no energy dissipation through wave breaking or friction at the seabed, no energy is imparted to the waves such as through the wind, and there is no seaward component of wave energy transfer through wave reflection.
- ***Small seabed slope***. The seabed slope is considered sufficiently small so that wave reflections do not occur and linear wave theory applies locally at any depth.

In developing the shoaling relations, intermediate depth conditions are related to those in deep water. Also, recall that $c_o = gT/2\pi$, $L_o = gT^2/2\pi$, and $k_o = 4\pi^2/gT^2$.

3.1.2 Shoaling Relations

Based on the above assumptions, the shoaling relations may be developed as follows:

$$\frac{c}{c_o} = \frac{L}{L_o} = \frac{k_o}{k} = \tanh(kd)$$

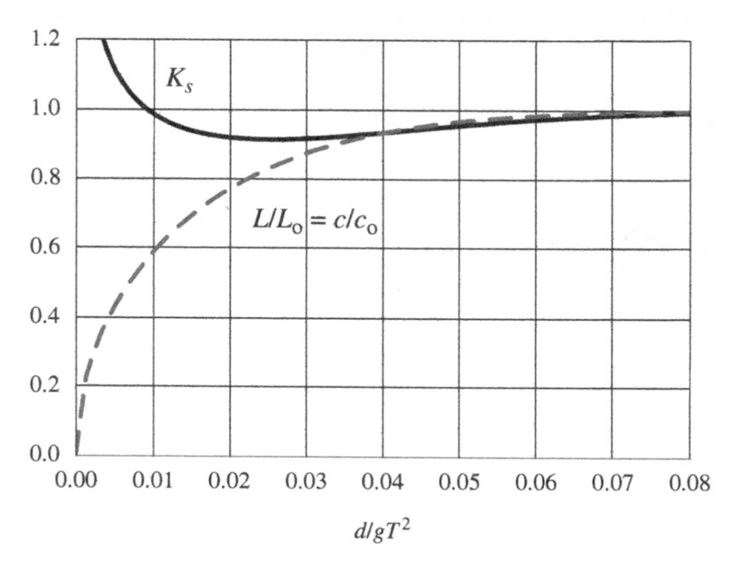

Figure 3.2 Wave shoaling relationships.

$$K_s = \sqrt{\frac{c_{Go}}{c_G}} = \sqrt{\frac{2\cosh^2(kd)}{2kd + \sinh(2kd)}}$$

where K_s is the *shoaling coefficient* defined as H_{sh}/H_o and H_{sh} denotes the wave height due to shoaling alone. Figure 3.2 shows the variation of c/c_o, L/L_o, and K_s with decreasing water depth as the waves propagate into shallower water, indicating how the wave length and wave speed reduce continually with depth, whereas the wave height initially reduces then increases as the waves propagate shoreward from deep water. The latter occurs on account of the non-monotonic variation of energy flux P with depth for a given wave height and period.

3.2 Wave Refraction

Wave refraction refers to changes to a wave train associated with waves approaching a shoreline obliquely, so that different portions of a wave crest encounter different water depths. Because of this, the portion of a wave crest in deeper water travels at a greater speed than the portion in shallower water, resulting in a bending of the wave crests so as to become more nearly parallel to the seabed contours. Figure 3.3 shows a photograph of wave refraction, illustrating how the wave crests become more closely parallel to the shore as they reach the shore.

3.2.1 Refraction Relations

Figure 3.4 provides a definition sketch of the fundamental situation to be analyzed. The figure shows a series of wave crests approaching shallow water, along with a complementary set of wave orthogonals.

Figure 3.3 Illustration of wave refraction. *Source:* NOAA/Public domain.

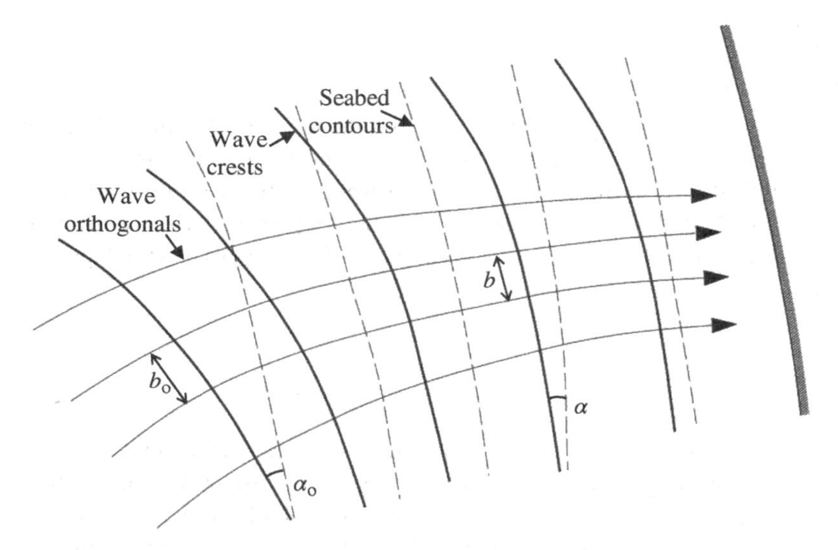

Figure 3.4 Definition sketch of wave refraction.

A *wave orthogonal*, sometimes referred to as a *wave ray*, is a line normal to wave crests and so coincides with the local wave direction. In the figure, α is the angle between the wave crests and seabed contours, b is the distance between a pair of adjacent orthogonals, and the subscript o represents conditions in deep water. Refraction relations are developed by expressing the wave height H and wave direction α at any one location in terms of the deep-water height H_o and deep-water direction α_o. (Note that changes in wave length are associated with wave shoaling only.)

As a wave crest propagates obliquely into shallow water, the portion in shallower water will move more slowly since wave speed reduces as the depth reduces. This differential speed along the crest leads to an associated change in wave direction. Thus, based on Snell's law and the dispersion relation, the wave direction α and the wave speed c may be shown to vary with the local depth parameter kd as follows:

$$\frac{\sin(\alpha)}{\sin(\alpha_o)} = \frac{c}{c_o} = \tanh(kd)$$

That is, the local direction α at any depth may be obtained from the deep-water wave direction α_o and the local depth parameter kd. Figure 3.5 may be used to estimate the local wave direction in terms of the alternate depth parameter d/gT^2 and the deep-water wave direction α_o.

Changes in wave height may be developed by assuming no energy dissipation, such that the product Pb is a constant, where P is the wave energy flux as described in Section 2.4 and b is the distance between a pair of orthogonals. Recognizing that shoaling accounts for changes in height due to changes in depth d and that refraction accounts for changes in height due to orthogonal spacing b, the wave height H due to both shoaling and refraction is given as:

$$\frac{H}{H_o} = K_s\, K_r$$

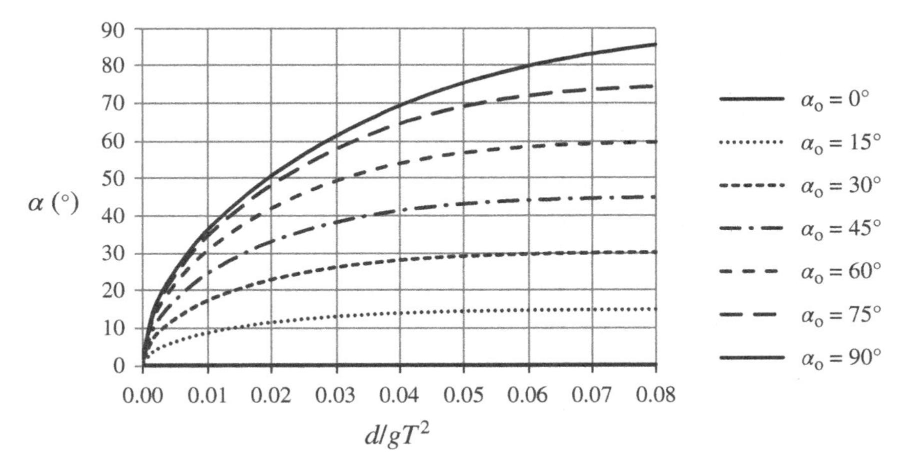

Figure 3.5 Wave direction changes due to wave refraction.

where K_s is the shoaling coefficient given in Section 3.1, and K_r is a *refraction coefficient* given by:

$$K_r = \sqrt{b_o/b}$$

For specified deep-water wave conditions and bathymetry, numerical techniques may be used to develop wave directions and wave heights over an area by projecting a set of orthogonals from deep water towards the shore.

Figure 3.6 illustrates the wave crests and orthogonals for two example cases. The first is a headland where the orthogonals are seen to converge, leading to higher wave heights adjacent to the headland. The second is a bay where the orthogonals are seen to diverge, leading to lower wave heights in the bay.

Figure 3.7 shows the reference case of a straight shoreline with straight parallel seabed contours, for which a closed-form solution for K_r is available. Because the pattern cannot change along the coastline

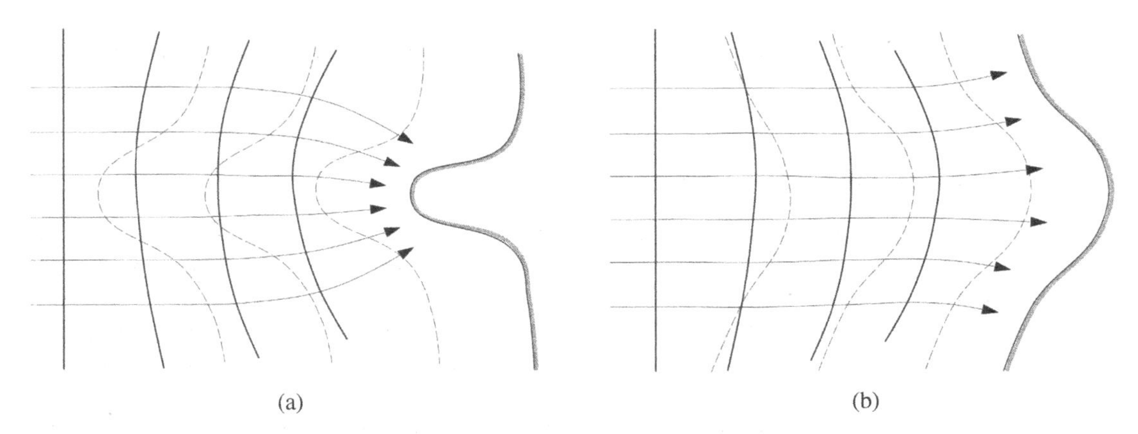

Figure 3.6 Examples of wave refraction. (a) Headland, (b) bay.

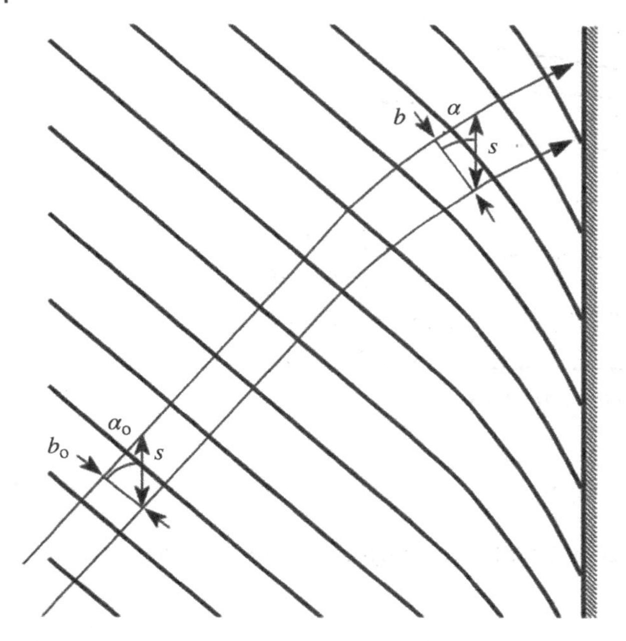

Figure 3.7 Definition sketch of wave refraction for a straight shoreline.

(i.e. the waves cannot bunch up), the distance s indicated in the figure must remain constant. Therefore:

$$\frac{b}{\cos\alpha} = \frac{b_o}{\cos\alpha_o}$$

This leads to a closed-form solution for K_r:

$$K_r = \left[\frac{1 - \sin^2\alpha_o \tanh^2(kd)}{\cos^2\alpha_o}\right]^{-1/4}$$

Figure 3.8 shows the refraction coefficient at various depths for different deep-water wave directions.

3.2.2 Numerical Modeling of Shoaling and Refraction

While the depiction of wave refraction using wave orthogonals is visually appealing, the use of wave orthogonals in a determination of wave heights at multiple locations over an area is cumbersome. Instead, a determination of wave heights and wave directions over a specified grid is more useful in analyzing wave conditions at a site. That is, rather than developing a numerical solution by marching orthogonals from deep water towards the shoreline, it is more convenient to develop wave height and direction results over a specified grid.

The following summary of such an approach is based on Dalrymple (1988). This is developed by recasting the shoaling/refraction relations as a pair of partial differential equations for the wave direction α

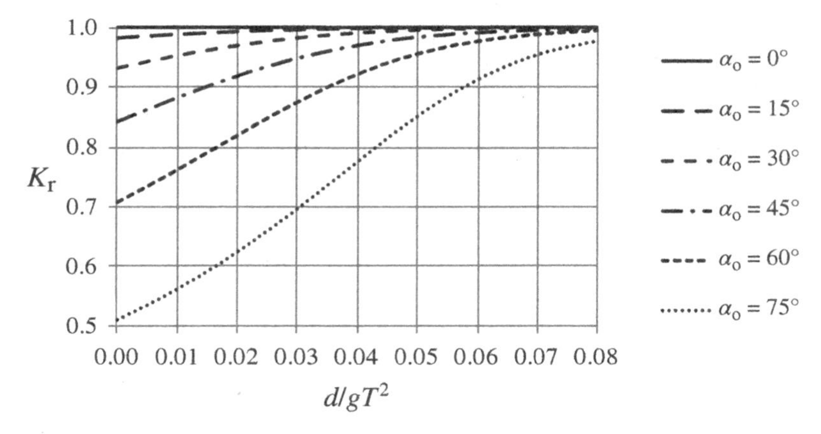

Figure 3.8 Refraction coefficient for a straight shoreline as a function of relative depth and deep-water wave direction.

(now defined with respect to a specified coordinate system rather than the seabed contours) arising from Snell's law and for the wave energy flux P arising from the conservation of wave energy. These take the following form:

$$\frac{\partial(k \sin \alpha)}{\partial x} - \frac{\partial(k \cos \alpha)}{\partial y} = 0$$

$$\frac{\partial(P \cos \alpha)}{\partial x} + \frac{\partial(P \sin \alpha)}{\partial y} = 0$$

where k is the wave number whose variation with location is known. These equations can be suitably recast with respect to finite difference modeling and thereby treated in conjunction with suitable boundary conditions so as to develop a solution for the general case.

Reference Solution A2 in Appendix A provides a spreadsheet solution to wave shoaling and refraction for specified deep-water wave conditions propagating over a domain with a specified bathymetry. The appendix lists various caveats to the use of this solution.

As an example of such a computation, Figure 3.9 shows results for waves approaching a shoal. Figure 3.9a indicates the grid that is used and the assumed bathymetry describing the shoal, while Figure 3.9b provides contours of the resulting wave heights and shows the wave directions. The results show the expected increase in wave heights over the shallower depths of the shoal and show wave directions converging towards the shoal.

While the model described here is somewhat basic, this outline is intended to provide a general indication of how high-level wave transformation models may be developed. As summarized in Section 3.11, such models can account for other forms of wave transformation as well, including wave diffraction, coexisting currents, wave energy dissipation, and wave breaking.

Example 3.1 ***Wave Shoaling and Refraction***

A wave train approaches a shoreline with straight, parallel seabed contours. At a location where the water depth $d = 4$ m, the wave period $T = 3.5$ s, the wave height $H = 1.1$ m, and the wave crests make an angle $\alpha = 60°$ with the seabed contours. What are the deep-water wave speed and direction? What are the shoaling coefficient and refraction coefficient? What is the deep-water wave height?

Solution

Specified parameters

$g = 9.80665\ \text{m/s}^2$
$H = 1.1\ \text{m}$
$T = 3.5\ \text{s}$
$d = 4.0\ \text{m}$
$\alpha = 60°$

Solve for *kd*

$d/gT^2 = 0.033$
$kd = 1.463$

Deep-water wave speed and direction

$L_o = gT^2/2\pi = 19.1\ \text{m}$
$c_o = L_o/T = 5.5\ \text{m/s}$

For α_o, the relevant formula is: $\sin(\alpha)/\sin(\alpha_o) = \tanh(kd)$. Therefore:

$$\alpha_o = \sin^{-1}\left[\frac{\sin(\alpha)}{\tanh(kd)}\right] = 74.6°$$

Shoaling and refraction coefficient

From the relevant formulae:

$$K_s = \sqrt{\frac{2\cosh^2(kd)}{2kd + \sinh(2kd)}} = 0.92$$

$$K_r = \left[\frac{1 - \sin^2\alpha_o \tanh^2(kd)}{\cos^2\alpha_o}\right]^{-1/4} = 0.73$$

Deep-water wave height

The relevant formula is $H/H_o = K_s\,K_r$. Therefore:

$$H_o = \frac{H}{K_s\,K_r} = 1.6\ \text{m}$$

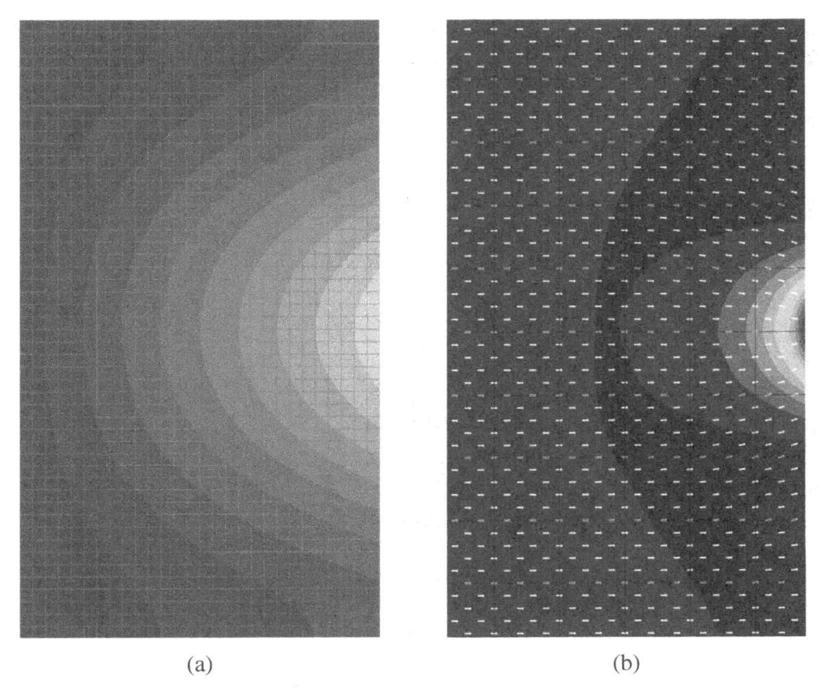

Figure 3.9 Example application of a wave shoaling/refraction numerical model. (a) Computational grid and bathymetry, (b) wave height contours and directions.

3.3 Wave Diffraction

Wave diffraction describes the propagation of waves around a vertical or other kind of barrier into a sheltered region. The case of constant depth is considered so that diffraction is treated in the absence of wave shoaling and refraction. Figure 3.10 provides an illustration of wave diffraction, showing wave crests diffracting around a series of offshore breakwaters with the crests bending so as to propagate into the sheltered area behind each breakwater.

3.3.1 Boundary Value Problem

The boundary value problem for wave diffraction may be developed by extending linear wave theory to three dimensions and accounting for the presence of a barrier in the flow. First, the definition of velocity potential ϕ may be extended to describe a flow in three dimensions as follows:

$$u = \frac{\partial \phi}{\partial x}; v = \frac{\partial \phi}{\partial y}; w = \frac{\partial \phi}{\partial z}$$

The three-dimensional Laplace equation now describes the flow throughout the fluid region:

$$\frac{\partial^2 \phi}{\partial x^2} + \frac{\partial^2 \phi}{\partial y^2} + \frac{\partial^2 \phi}{\partial z^2} = 0$$

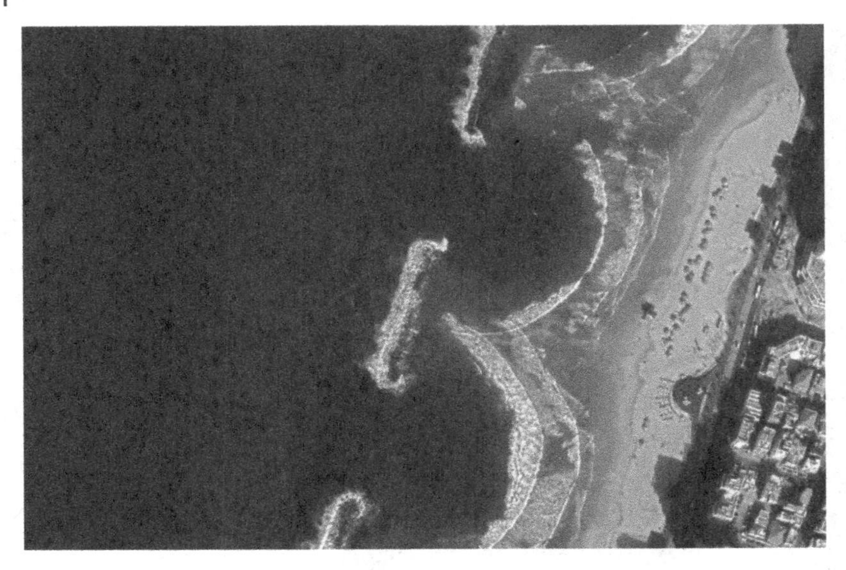

Figure 3.10 Illustration of wave diffraction around offshore breakwaters. *Source:* Google Earth.

Boundary conditions at the seabed and the still water level are unchanged. However, the presence of a barrier is accounted for by an additional boundary condition, which states that the flow velocity on the barrier surface in the normal direction is zero:

$$\frac{\partial \phi}{\partial n} = 0 \quad \text{at the barrier surface}$$

where n is distance normal to the surface.

In order to take account of this boundary condition, the velocity potential is considered to represent the superposition of two flows, one corresponding to the incident wave field, as if the barrier was absent, and an additional flow that coexists with the incident wave field so as to assure that this boundary condition is satisfied. Thus, ϕ is expressed as:

$$\phi = \phi_{\mathrm{W}} + \phi_{\mathrm{S}}$$

where ϕ_{W} is the incident potential known from linear wave theory, and ϕ_{S} is termed a scattered potential, corresponding to the correction flow that is to be determined from the boundary condition on the barrier surface. Thus, the barrier surface boundary condition is recast as:

$$\frac{\partial \phi_{\mathrm{S}}}{\partial n} = -\frac{\partial \phi_{\mathrm{W}}}{\partial n} \quad \text{at the barrier surface}$$

Since the problem is linear, the other governing equations, including the Laplace equation, the free surface boundary conditions, and the seabed boundary condition all apply separately to both ϕ_{S} and ϕ_{W}.

However, in order to ensure a unique solution for the scattered potential, ϕ_{S} is also required to satisfy an additional condition, termed the *radiation condition*, in the far field at some distance away from the barrier. This requires that the scattered waves are outgoing only and that the wave height decays appropriately with distance because of the directional spread of energy. Using a complex representation of the velocity potential, this takes the form:

$$\underset{r \to \infty}{\text{Limit}} \sqrt{r}\left(\frac{\partial \phi_{\mathrm{S}}}{\partial r} - ik\phi_{\mathrm{S}}\right) = 0$$

where r is radial distance measured from an origin in the vicinity of the barrier.

applied to the inner breakwater, such that the incident wave train with height H_B at the breakwater tip B, now approaching in the different direction shown, leads to the diffracted wave height at point C. In utilizing this approximation, care should be taken with respect to the possibility of multiple reflections between the overlapping breakwaters.

Breakwater orientation. The solution is not strongly dependent on breakwater orientation. Because of this, the breakwater may be rotated about its tip, either clockwise or counterclockwise, without rotating the incident wave direction and without rotating the points of interest in the sheltered region. This makes it possible to apply a diffraction diagram with a standard wave direction to the case of a nonstandard wave direction. More formally, if the incident wave direction θ_o and the direction θ of the required location are specified, then one would use a diffraction diagram with an effective incident direction θ'_o corresponding to a standard direction, so as to determine the diffraction coefficient along an effective direction $\theta' = \theta + \theta'_o - \theta_o$. Such a case is illustrated in Example 3.2.

Example 3.2 ***Wave Diffraction Around a Breakwater***

What is the wave height at point A in the figure below?

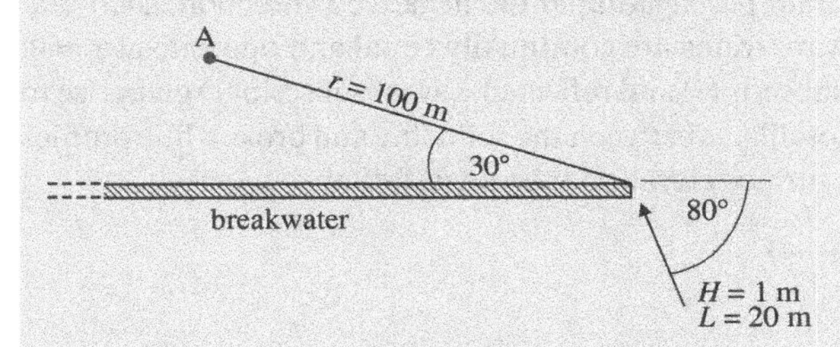

Solution

Specified parameters

$H = 1.0\,\text{m}$
$L = 20.0\,\text{m}$
$r = 100.0\,\text{m}$
$\theta_o = 80°$
$\theta = 30°$

Method A – based on spreadsheet solution (see Reference Solution A3)

Apply the above parameters to the Reference Solution A3 spreadsheet. This gives $K_d = 0.128$. Therefore:

$$H_A = H\,K_d = 0.13\,\text{m}$$

Method B – based on diffraction diagrams (see Reference Solution A4)

Select the diagram with a direction θ'_o that is closest to θ_o. This is $\theta'_o = 75°$.

Use a mirror image of the diagram with the following parameters:

$r/L = 5$
$\theta' = \theta + \theta'_o - \theta_o = 25°$
Reading off from diagram, $K_d = 0.13$. Therefore:
$H_A = HK_d = 0.13\,\text{m}$

3.4 Standing Waves

3.4.1 Standing Waves at a Wall

Consideration is now given to a normally incident wave train that reflects from a vertical wall, so as to give rise to standing waves as sketched in Figure 3.15.

Standing waves develop when an incident wave train propagates in the x direction and encounters a vertical wall at $x = 0$, as indicated in the figure. The presence of the wall introduces an additional boundary condition requiring that the horizontal velocity at the wall is zero. This boundary condition is satisfied by the introduction of a reflected wave train propagating in the negative x direction, such that the horizontal velocities of the two component wave trains are continually equal and opposite at $x = 0$. The combined wave field due to incident waves (subscript i) and reflected waves (subscript r) gives rise to standing waves, with a free surface elevation that oscillates between the solid line and broken line profiles shown in Figure 3.15. Thus, the combined water surface elevation is given as follows:

$$\eta = \eta_i + \eta_r = \frac{H}{2}\cos(kx - \omega t) + \frac{H}{2}\cos(kx + \omega t)$$
$$= H\cos(kx)\cos(\omega t)$$

The combined wave field does not propagate, but rather appears as "standing" such that, at certain locations the water surface elevation oscillates between crests and troughs, but the crests do not move horizontally. Specifically, the surface elevation is continually zero at the nodes, which are at $x = \pm L/4$, $\pm 3L/4, \dots$, and the surface elevation oscillates vertically between crests and troughs at the antinodes, which are at $x = 0, \pm L/2, \pm L, \dots$

Formulae for the various flow parameters for standing waves are given in Table 3.1. Note that H in the table denotes the wave height of the incident waves so that the standing wave height is $2H$. It may be confirmed that the above formula for u shows that the horizontal velocity is continually zero at the antinodes, which include the wall, and so satisfies the boundary condition at the wall.

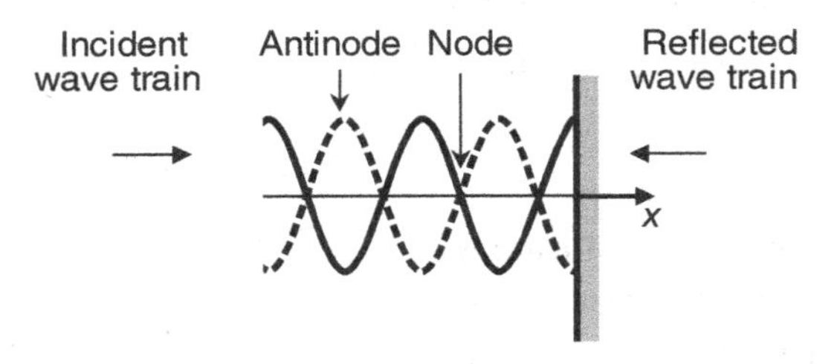

Figure 3.15 Definition sketch of standing waves.

Table 3.1 Results for linear standing waves.

Variable	Equation
Free surface elevation	$\eta = \mathrm{H}\cos(kx)\cos(\omega t)$
Velocity potential	$\phi = \dfrac{2\pi H}{kT}\dfrac{\cosh(ks)}{\sinh(kd)}\cos(kx)\sin(\omega t)$
Dispersion relation	$\omega^2 = gk\tanh(kd)$
Horizontal displacement	$\xi = -H\dfrac{\cosh(ks)}{\sinh(kd)}\sin(kx)\cos(\omega t)$
Vertical displacement	$\zeta = H\dfrac{\sinh(ks)}{\sinh(kd)}\cos(kx)\cos(\omega t)$
Horizontal velocity	$u = \dfrac{2\pi H}{T}\dfrac{\cosh(ks)}{\sinh(kd)}\sin(kx)\sin(\omega t)$
Vertical velocity	$w = -\dfrac{2\pi H}{T}\dfrac{\sinh(ks)}{\sinh(kd)}\cos(kx)\sin(\omega t)$
Horizontal acceleration	$\dot{u} = \dfrac{4\pi^2 H}{T^2}\dfrac{\cosh(ks)}{\sinh(kd)}\sin(kx)\cos(\omega t)$
Vertical acceleration	$\dot{w} = -\dfrac{4\pi^2 H}{T^2}\dfrac{\sinh(ks)}{\sinh(kd)}\cos(kx)\cos(\omega t)$
Pressure	$p = -\rho g z + \rho g H\dfrac{\cosh(ks)}{\cosh(kd)}\cos(kx)\cos(\omega t)$

3.4.2 Standing Waves in a Basin

The features of a standing wave train may be exploited so as to develop standing waves in closed and open-ended basins. Since the horizontal velocity is zero at an antinode, a second vertical wall may be inserted at any antinode so that this flow corresponds to standing waves in a closed basin such as a lake or reservoir. Likewise, since the dynamic pressure is continually zero at a node, this location may be taken to correspond to the end of a narrow basin open to the ocean, where the pressure is approximately hydrostatic only, so that this flow corresponds to standing wave oscillations in an open-ended basin such as a narrow harbor open to the ocean.

Based on these approximations, Figure 3.16 depicts the water surface elevation corresponding to the lowest modes of standing waves in closed and open-ended basins. The modes refer to successive wave patterns with reducing wave length.

The figure also indicates the wave length L for the various modes in terms of the basin length ℓ. Based on an inspection of the wave patterns in the figure, it is possible to express the wave length L_n for the nth mode in terms of the basin length ℓ as follows:

$$L_\mathrm{n} = \begin{cases} \dfrac{2\ell}{n} & \text{for a closed basin} \\ \dfrac{4\ell}{(2n-1)} & \text{for an open-ended basin} \end{cases}$$

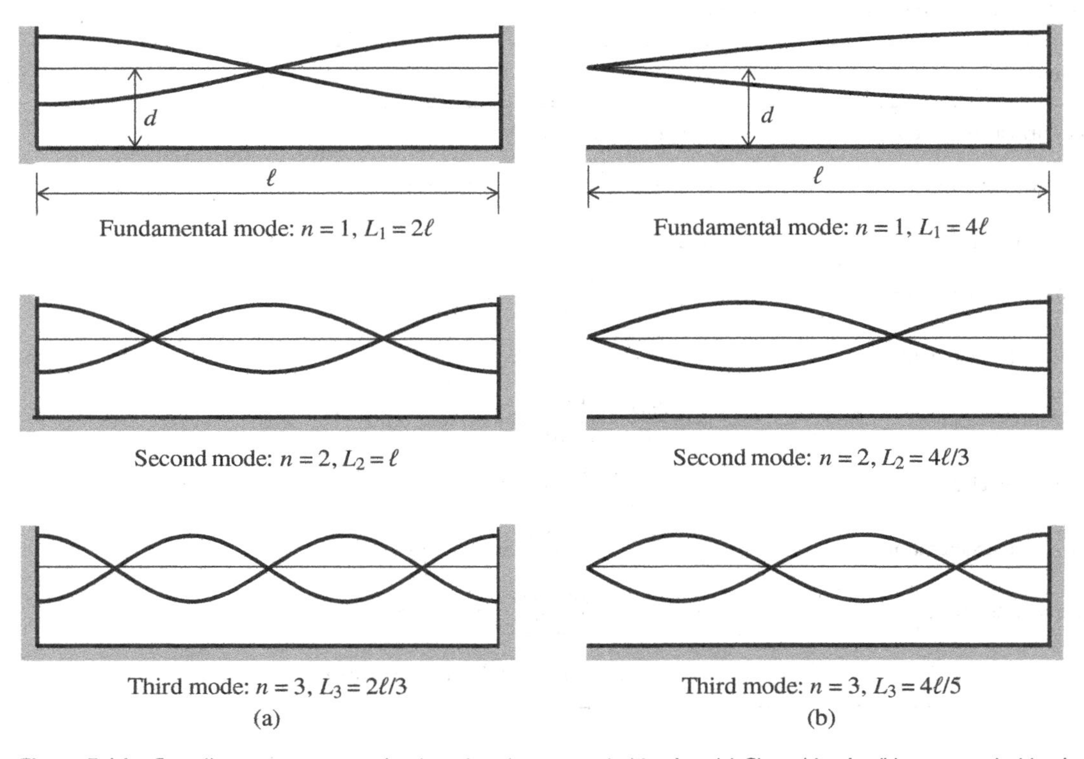

Figure 3.16 Standing wave systems in closed and open-ended basins. (a) Closed basin, (b) open-ended basin.

If $\ell \gg d$, then these are shallow waves for which $c = L/T = \sqrt{gd}$, so that the above formula can then be used to develop oscillation periods as $T_{\mathrm{n}} = L_{\mathrm{n}}/\sqrt{gd}$.

Standing waves in closed and open-ended basins do not usually apply to wind-generated waves but more commonly apply to sloshing in water tanks and reservoirs and to large-period oscillations in lakes and harbors termed seiche (see Section 7.4). It is noted that these standing waves require an initiation mechanism in order to develop, as arising, for example, from an earthquake, a tsunami, or storm surge.

Example 3.3 ***Standing Waves***

A wave train with period $T = 3.5$ s and a water depth $d = 4$ m undergoes perfect reflection at a vertical seawall. The wave crest elevation at the wall is 0.5 m. What is the minimum pressure at the base of the wall? What is the maximum horizontal velocity at the SWL 5 m from the wall? How far in front of the wall is the horizontal acceleration a maximum? For what length of an open-ended basin do these waves represent second-mode standing wave oscillations? Assume that $\rho = 1025\,\mathrm{kg/m^3}$.

Solution

Specified parameters

$g = 9.80665\,\mathrm{m/s^2}$
$\rho = 1025\,\mathrm{kg/m^3}$

$d = 4.0\,\mathrm{m}$
$T = 3.5\,\mathrm{s}$
η_c (at $x = 0$) $= 0.5\,\mathrm{m}$

Wave height and wave length

H (standing) $= 2 \times \eta_c = 1.0\,\mathrm{m}$
H (incident) $= 0.5 \times H$ (standing) $= 0.5\,\mathrm{m}$
$d/gT^2 = 0.033$
$kd = 1.463$
$k = kd/d = 0.366$
$L = 2\pi/k = 17.2\,\mathrm{m}$

Pressure, velocity, and acceleration

The required minimum pressure at the base of the wall ($kx = 0$, $z = -d$) occurs when $\cos(\omega t) = -1$. Thus, from Table 3.1:

$$p = -\rho g d - \rho g H \left[\frac{1}{\cosh(kd)}\right] = 38\,\mathrm{kPa}$$

The maximum horizontal velocity u_m at $x = -5\,\mathrm{m}$, $s = d$ is given by:

$${}_{\mathrm{m}} = \frac{2\pi H}{T}\frac{\cosh(ks)}{\sinh(kd)}\sin(kx) = 1.0\,\mathrm{m/s}$$

The horizontal acceleration is a maximum when $\sin(kx) = 1$. Hence, the lowest negative x value in front of wall is given by:

$kx = -\pi/2$
$x = -\pi/2k = -4.3\,\mathrm{m}$

Basin Length

For second-mode oscillations in an open-ended basin, $L = 4\ell/3$. Therefore:
$\ell = 3L/4 = 12.9\,\mathrm{m}$

3.5 Wave Reflection

3.5.1 Normal Reflection

The case of perfect reflection of a normally incident wave train at a vertical wall can be extended to the partial refection of a wave train, with the wave height of the reflected wave train now lower than that of the incident wave train. Figure 3.17 depicts the surface elevations of the incident and reflected wave trains with respect to a partially reflecting wall or barrier.

Since the reflected wave train propagates in the negative x direction, the surface elevation of the incident and reflected wave trains are given respectively as:

$$\eta_i = \frac{H_i}{2}\cos(kx - \omega t)$$

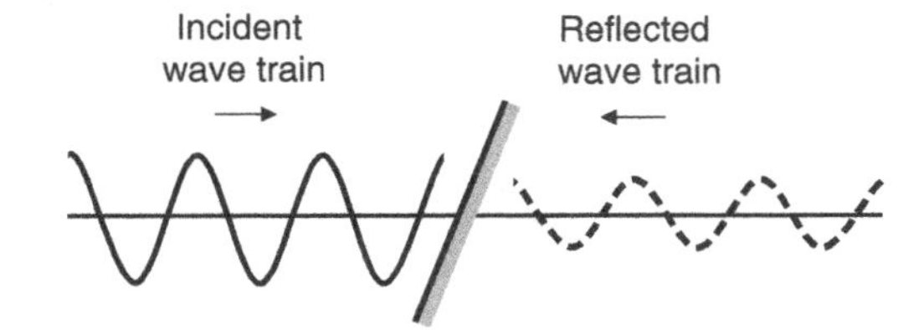

Figure 3.17 Sketch of partial wave reflection.

$$\eta_r = \frac{H_r}{2} \cos(kx + \omega t)$$

where H_i and H_r are the wave heights of the incident and reflected wave trains, respectively. The combined surface elevation is:

$$\eta = \eta_i + \eta_r = \frac{H_i}{2} [\cos(kx - \omega t) + K_r \cos(kx + \omega t)]$$

where K_r is a *reflection coefficient* defined as $K_r = H_r/H_i$. The reflection coefficient depends on the nature and slope of the wall or barrier. Typical ranges of K_r include the following:

$$K_r = \begin{cases} \approx 1 & \text{for a vertical wall} \\ < 1 & \text{for a sloping wall or rubble-mound structure} \\ < 0.1 & \text{for a beach} \end{cases}$$

The Coastal Engineering Manual (2002) provides information on reflection coefficient values for a range of shoreline structures.

3.5.2 Oblique Reflection

The preceding description refers to normally incident waves. When waves approach a wall obliquely, the resulting oblique wave reflection results in a diamond-shaped pattern of intersecting wave crests and troughs as indicated in Figure 3.18. Assuming perfect wave reflection ($K_r = 1$) for this case, the surface elevations of the incident and reflected wave trains are given as:

$$\eta_i = \frac{H}{2} \cos(kx \cos\alpha + ky \sin\alpha - \omega t)$$

$$\eta_r = \frac{H}{2} \cos(-kx \cos\alpha + ky \sin\alpha - \omega t)$$

where y is distance parallel to the wall. These may be combined to give:

$$\eta = \eta_i + \eta_r = H \cos(kx \cos\alpha) \cos(ky \sin\alpha - \omega t)$$

This equation corresponds to a diamond pattern of short-crested waves with three-dimensional crests and troughs that propagate parallel to the wall at a wave speed $c_y = c/\sin\alpha$.

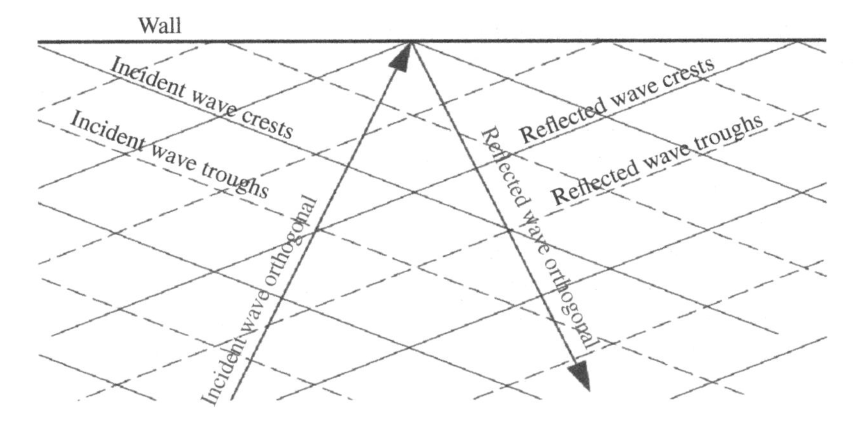

Figure 3.18 Sketch of oblique wave reflection.

3.6 Wave Transmission

In some cases, waves may also be transmitted past a barrier. Examples include waves propagating past a rubble-mound breakwater through overtopping and/or percolation through the breakwater, waves propagating over a submerged breakwater, waves propagating beneath a pile-supported vertical barrier that incorporates a gap adjacent to the seabed, and waves propagating past a floating breakwater, such that the transmitted waves may be influenced by the breakwater motions.

The general situation is indicated in Figure 3.19, which shows the incident, reflected and transmitted wave trains adjacent to such a barrier. Apart from the energy flux associated with each of these, energy may also be dissipated (or extracted in the case of a wave energy device) as depicted in the figure.

Given that energy flux is proportional to wave height squared, a balance of energy fluxes for the general case leads to the following equation:

$$K_r^2 + K_t^2 + K_e = 1$$

where K_r is the reflection coefficient, $K_t = H_t/H_i$ is a *transmission coefficient*, K_e is the proportion of incident wave energy flux that is dissipated or extracted, H_i is the incident wave height, and H_t is the transmitted wave height.

Further information on the transmission coefficient for floating breakwaters is provided in Section 8.8.1. For the present, consideration is given to the fundamental case of a thin vertical

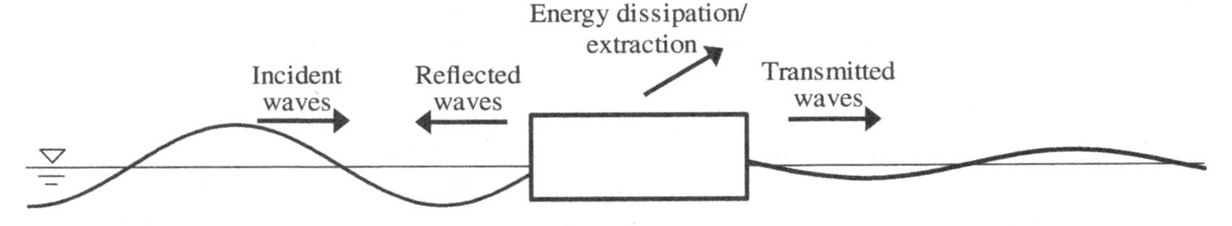

Figure 3.19 Depiction of wave transmission past a barrier.

barrier of draft h in water of depth d, so that the gap beneath the barrier has a height $w = d - h$. It is of interest to determine K_t as a function of h/L for various values of h/d.

A solution for the corresponding potential flow may be obtained by developing suitable representations of the flow potentials on the upwave and downwave sides of the barrier and applying the boundary condition of zero horizontal velocity at the barrier surface, along with matching conditions for the velocity and pressure distributions across the gap.

Apart from the transmitted waves that propagate downwave of the breakwater, the resulting flow field contains also a series of evanescent waves on either side of the barrier. These are standing waves whose amplitudes diminish with distance from the barrier, so that they do not impact the transmission coefficient. They are associated with the imposition of the boundary condition along the barrier sides, which cannot be satisfied by the incident, reflected and transmitted wave trains alone. Section 6.5.2 describes a different solution containing evanescent waves, in the context of the wave field adjacent to a wave generator in a laboratory flume.

A much simpler, yet reliable approximation for the transmission coefficient has been developed by Kriebel and Bollmann (1996) on the basis of energy considerations. Their solution for the transmission coefficient is given as:

$$K_t = \frac{2T_F}{1 + T_F}$$

where:

$$T_F = \frac{2kw + \sinh(2kw)}{2kd + \sinh(2kd)}$$

Although this solution breaks down in the long-wave limit (for which $K_t = w/d$ rather than $K_t = 1$ is obtained), it otherwise appears to provide a better fit to experimental data than does the complete solution.

3.7 Wave Attenuation

3.7.1 Forms of Energy Dissipation

Energy dissipation in waves may occur in various ways and leads to an attenuation of a wave train. Dissipation occurs most significantly as waves break when reaching a shore – the behavior of breaking waves and broken waves will be considered shortly. Prior to wave breaking, wave energy is dissipated in intermediate and shallow depths by friction and percolation at the seabed. Related forms of dissipation are associated with vegetative shorelines and muddy bottoms. Also, as waves are generated by the wind, energy dissipation arises on account of wind-induced wave breaking in the form of whitecaps. Finally, under laboratory conditions, energy is dissipated by friction at the bottom and side-walls of a flume, and, if scales are very small, by surface tension effects on the water surface.

Of these various forms of energy dissipation, the one associated with friction on the seabed is of particular interest. In order to explore this, a parameter D is defined as the average rate of energy dissipation per unit length in the wave direction and per unit width and may be expressed as $D = \overline{\tau u}$, where τ is the time-varying shear stress at the seabed, u is the fluid velocity at the outer edge of the seabed boundary layer, and the overbar denotes a temporal mean. If both τ and u are sinusoidal with amplitudes τ_w and u_o, respectively, and δ is the phase angle between them, then D may be expressed as $D = ½\tau_w u_o \cos\delta$. The fluid velocity is known from linear wave theory, so that information on τ_w and δ are needed in order to determine D. τ_w is commonly expressed in dimensionless form as a *friction factor*, defined as $f_w = \tau_w / ½\rho u_o^2$. Information on f_w is described below.

3.7.2 Friction Factor

The friction factor f_w is taken to depend on the wave Reynolds number $\mathrm{Re}_w = u_o a/\nu$, where $a = d_o/2$ is the orbital amplitude of the fluid displacement at the seabed and ν is the kinematic viscosity of water, and/or a relative roughness parameter a/k_s, where k_s is a characteristic roughness size. The behavior of f_w depends on the boundary-layer flow regime at the seabed, which may be considered to correspond to either a laminar boundary layer, a smooth, turbulent boundary layer, or a rough, turbulent boundary layer. A number of experimental studies, including that reported on by Kamphuis (1975), and theoretical analyses have been carried out in order to investigate the dependence of f_w on Re_w and a/k_s. For a laminar boundary layer, f_w may be obtained on a theoretical basis in a fairly straightforward manner. For a smooth, turbulent boundary layer, different studies have proposed slightly different equations expressing f_w in terms of Re_w, while, for a rough, turbulent boundary layer, they have proposed slightly different equations expressing f_w in terms of a/k_s. Overall, suitable equations for the three flow regimes may be taken as follows:

Laminar flow:

$$f_w = \frac{2}{\sqrt{\mathrm{Re}_w}}$$

Smooth, turbulent flow:

$$\frac{1}{8.1\sqrt{f_w}} + \log_{10}\left(\frac{1}{\sqrt{f_w}}\right) = -0.135 + \log_{10}(\sqrt{\mathrm{Re}_w})$$

Rough, turbulent flow:

$$\frac{1}{4\sqrt{f_w}} + \log_{10}\left(\frac{1}{4\sqrt{f_w}}\right) = -0.08 + \log_{10}\left(\frac{a}{k_s}\right)$$

The above equations are plotted in Figure 3.20 in the form of a Moody diagram that has traditionally been used for the case of a steady flow. This shows f_w as a function of Re_w for the laminar and smooth, turbulent flow regimes and as independent of Re_w for various values of a/k_s for the rough, turbulent flow regime.

In order to apply the above equations, an estimate of the roughness height is needed. For a flat seabed, this depends on sediment grain size, with one estimate given as $k_s = 3D$, where D is the median diameter of the grains. However, k_s may be influenced by the presence of bedforms, usually in the form of ripples, and by other forms of roughness, such as the presence of seagrasses on the seabed.

In applying the above equations, the Re_w values at which the flow transitions from one flow regime to another are also needed. In particular, the transition from a laminar flow to a smooth, turbulent flow occurs when $\mathrm{Re}_w \simeq 10^5$, while the transition from a laminar flow to a rough, turbulent flow occurs over

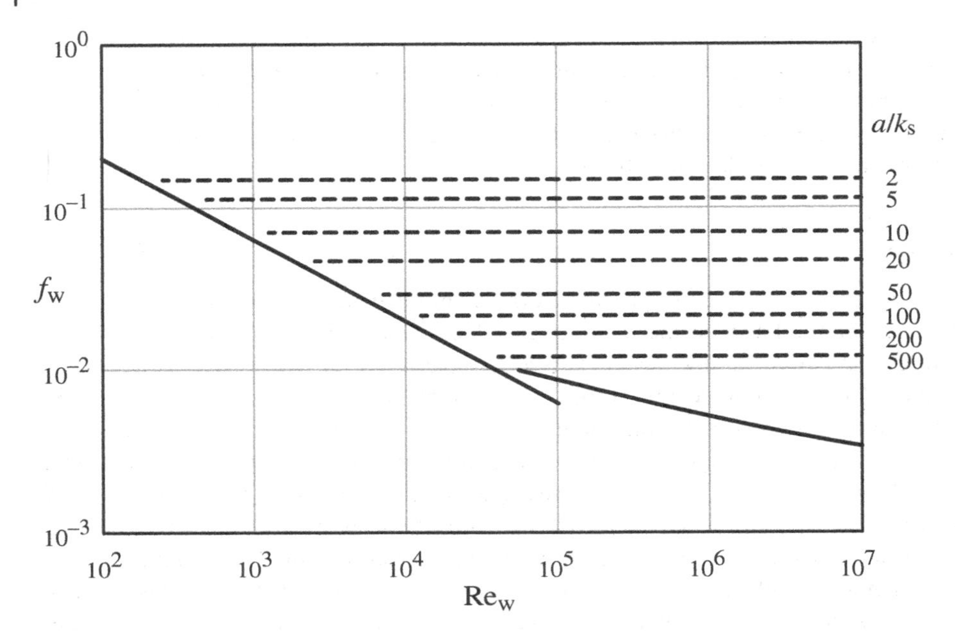

Figure 3.20 Friction factor as a function of wave Reynolds number Re_w and relative roughness a/k_s.

a range of Re_w, depending on the value of a/k_s. However, a simple, practical approach is to adopt the maximum value of f_w as given by each of the three expressions.

3.7.3 Attenuation Rate

Once f_w, δ, and therefore D are known for particular conditions, the extent of wave attenuation described by the rate of decrease of wave height H with distance, $\mathrm{d}H/\mathrm{d}x$, may be assessed. For the two-dimensional case, this may be developed by equating the difference in average rates of energy flux crossing planes a short distance $\mathrm{d}x$ apart to the average rate of energy dissipation between those planes; that is by taking $\mathrm{d}P/\mathrm{d}x = -D$. Such an approach may be used to develop an estimate of $\mathrm{d}H/\mathrm{d}x$. It is thereby found that, for the case of a turbulent boundary layer on the seabed, the wave height attenuates such that $\mathrm{d}H/\mathrm{d}x$ is proportional to $-H^2$, with different relationships obtained for various cases. As one example, information on the degree of attenuation of waves propagating in a channel with rubble-mound sides is available.

More generally, information on f_w may be applied to numerical wave transformation models in order to take account of friction at the seabed in assessing changes to waves. Such models may be extended to take account of the effects of vegetation, a muddy bottom, and rough channel slopes.

3.8 Waves of Maximum Height

A wave train propagating over a horizontal seabed has a maximum possible wave height for a given wave length and water depth. As waves approach such a limit, they become unstable and break. Thus it is of interest to establish the maximum possible wave height for a progressive wave train in water of constant depth. In deep water, the maximum wave height H_m is limited by the *wave steepness*, which is the wave-height-to-wave-length ratio. In shallow water, it is limited by the wave-height-to-depth ratio. On the basis of theoretical and numerical studies, these limits have been found to be as follows:

In deep water: $H_m/L = 0.141$
In shallow water: $H_m/d = 0.827$

A common formula for the maximum wave height applicable at all depths is as follows:

$$H_m/L = 0.142\ \tanh(kd)$$

However, this simplified formula overpredicts the maximum wave height in shallow water, giving $H_m/d = 0.89$ in lieu of 0.83. It is emphasized that the above description refers only to progressive waves in water of constant depth.

3.9 Breaking Waves

Waves usually break as they approach a shoreline, as the water depth reduces and the wave height increases. The *surf zone* refers to the zone adjacent to a shoreline within which wave breaking occurs. In distant or aerial views of the shoreline, this often shows as a white band adjacent to the shoreline, as indicated in Figure 3.21. (Waves may also break in the open ocean in the form of whitecaps caused by the wind, but such a case is not considered here.)

Figure 3.21 View of the surf zone, Honolulu, HI.

For waves breaking near a shoreline, the form of wave breaking, the breaking wave height H_b, and the depth at breaking d_b are of primary interest. These are found to depend principally on the *surf similarity parameter* ξ that is defined as:

$$\xi = \frac{m}{\sqrt{H_o/L_o}}$$

where $m = \tan\beta$ is the beach slope, β is the angle of the seabed relative to the horizontal, and H_o/L_o is the deep-water wave steepness.

3.9.1 Forms of Wave Breaking

The forms of wave breaking are classified as spilling, plunging, and surging breakers, as illustrated in Figures 3.22 and 3.23.

Spilling, plunging, and surging breakers tend to occur on gently sloped beaches, steeper beaches, and very steep beaches, respectively. More particularly, summary descriptions of the three forms of breaking waves, and the corresponding ranges of the surf similarity parameter under which they occur, as described in the Coastal Engineering Manual (2002), are summarized in Table 3.2.

3.9.2 Breaking Wave Height and Depth

Estimates of the breaking wave height H_b and the depth at breaking d_b are of interest. H_b and d_b are often expressed in terms of the *breaker index* $\gamma_b = H_b/d_b$. A simple estimate of the breaker index independent of beach slope is given as:

$$\frac{H_b}{d_b} = 0.78$$

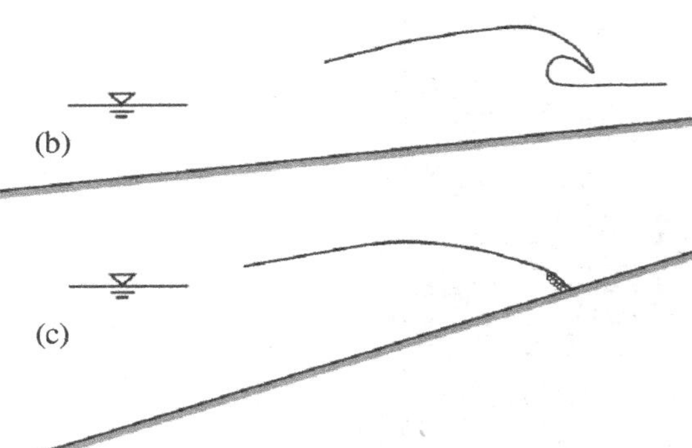

Figure 3.22 Illustration of types of breaking waves. (a) Spilling breaker, (b) plunging breaker, (c) surging breaker.

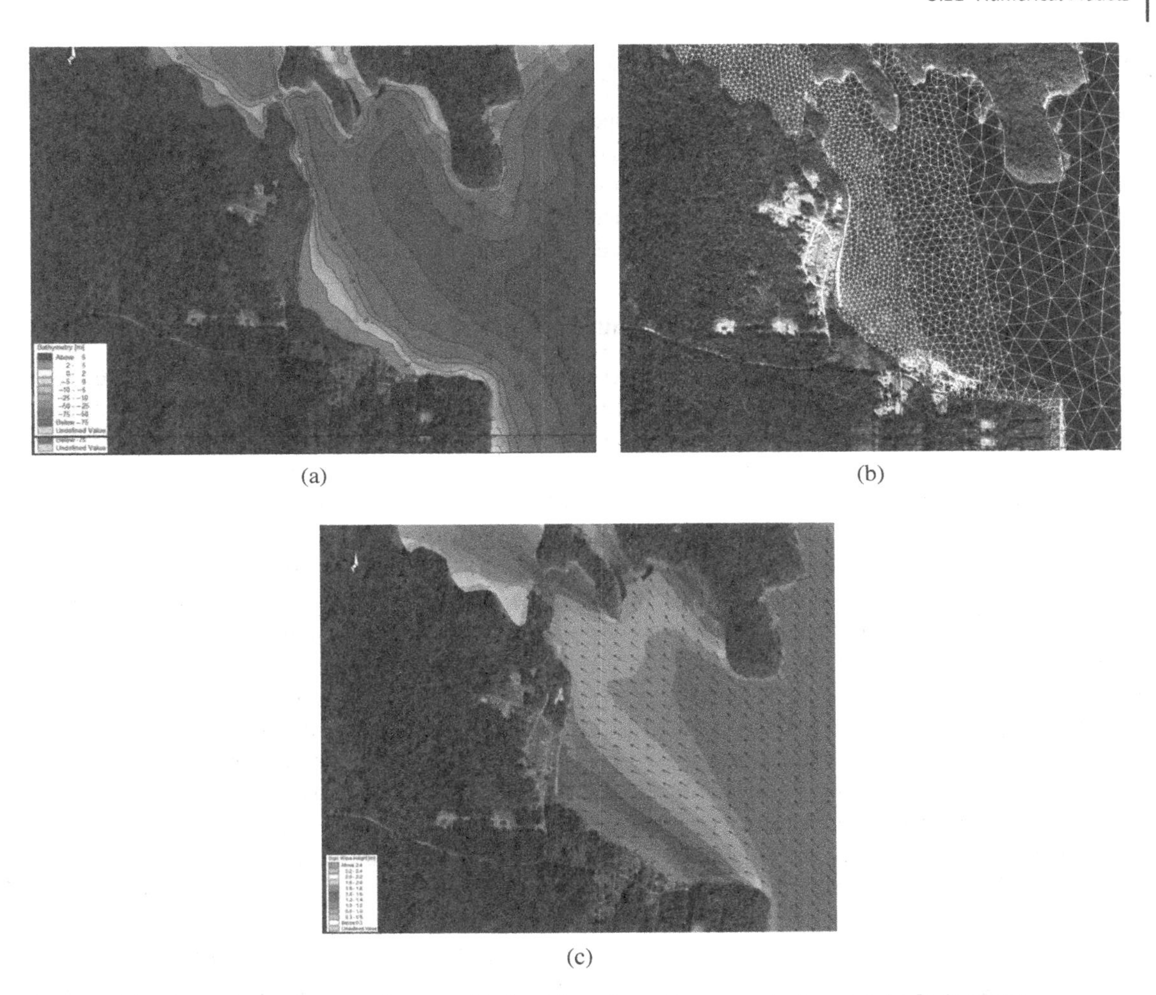

Figure 3.25 Example application of a wave hindcast/transformation numerical model. (a) Bathymetry, (b) computational grid, (c) wave height contours and directions. *Source:* Reproduced with permission of CMO Consultants Ltd.

where $A = -igH/2\omega$, and $f(x, y)$ is a two-dimensional function in the horizontal plane whose solution is to be determined. That is, ϕ is represented by a hyperbolic cosine variation with depth, so that the three-dimensional description of ϕ is reduced to a two-dimensional problem for $f(x, y)$ in the horizontal plane. The above representation satisfies the seabed and free surface boundary conditions, while the governing equation for $f(x, y)$ is found to be:

$$\frac{\partial}{\partial x}\left(cc_{\mathrm{G}}\frac{\partial f}{\partial x}\right) + \frac{\partial}{\partial y}\left(cc_{\mathrm{G}}\frac{\partial f}{\partial y}\right) + k^2 cc_{\mathrm{G}} f = 0$$

where c is the wave speed, c_G is the group velocity, and k is the wave number. These all depend on depth and so vary with location (x, y). Taking account of suitable boundary conditions in the horizontal plane, this governing equation, referred to as a mild-slope equation, is then solved numerically for f.

Modified and extended forms of the mild-slope equation have been proposed in order to take account of, for instance, steeper seabed slopes, wave–current interactions, seabed friction, wave breaking, and wave nonlinearities. In cases where wave reflection is negligible, a "parabolic approximation" to the mild-slope equation can be developed so as to lead to a considerable reduction in computational effort.

3.11.3 Models Based on Boussinesq-Type Equations

Recall that nonlinear shallow-water wave theories (referred to in Section 2.7 as the "cnoidal wave theories") may be developed by initially "stretching" the vertical length scale relative to the horizontal length scale, along with a perturbation series representation of the flow and Taylor series expansions to treat the free surface boundary conditions. This type of treatment of the nonlinear shallow wave equations may be extended so as to incorporate variable depths, nonperiodic time variations, and other forms of wave transformation, thereby extending greatly their range of application.

The general approach stems from the work of Boussinesq in 1872, developed further by Peregrine (1967) with respect to variable depths and by many other authors since the 1990s. Since a range of formulations have been developed on the basis of different assumptions and approximations, such methods are referred to as being based on Boussinesq-type equations.

There are two key aspects to the development of such approaches. First, suitable account is taken of nonlinear effects, characterized by a representative amplitude-to-depth ratio and denoted ε, and of frequency dispersion effects (whereby wave speed depends on wave frequency), characterized by a representative depth-to-wavelength ratio and denoted μ. Analogous to the scaling or "stretching" referred to above, different approximations reliant on ε and μ may be made in different ways so as to develop different versions of the continuity and momentum conservation equations for an incompressible fluid and an irrotational flow. The representation of frequency dispersion effects characterized by μ has now improved to the extent that these methods are applicable to relative depths that approach deep-water conditions.

Second, the vertical variation of parameters is approximated so as to reduce the three-dimensional problem to a two-dimensional problem in the horizontal plane. In particular, the vertical variation of horizontal velocity is commonly taken as quadratic, with the vertical variation of vertical velocity then being linear.

By way of illustration, when the horizontal velocity is represented by the depth-averaged velocity, denoted U, and only the x direction (with varying depth) is considered, the particular form of governing equations developed by Peregrine (1967) may be written as:

$$\frac{\partial \eta}{\partial t} + \frac{\partial}{\partial x}[(d + \eta)U] = 0$$

$$\frac{\partial U}{\partial t} + g\frac{\partial \eta}{\partial x} + U\frac{\partial U}{\partial x} = \frac{d}{2}\frac{\partial}{\partial t}\left(\frac{\partial^2 U}{\partial^2 x}\right) - \frac{d^2}{6}\frac{\partial^2 U}{\partial^2 x}$$

where $d(x)$ is the depth and $U(x, t)$ is the depth-averaged horizontal velocity. A brief discussion of the application of the Boussinesq-type equations to the development of long wave numerical models is provided in Section 7.1.2.

←■

Problems

3.1 Waves approach normally a beach with straight parallel bottom contours. Measurements at a location where the depth $d = 8$ m indicate that the wave height $H = 1.5$ m and the period $T = 7.0$ s. What are the deep-water wave length and the deep-water wave height? What is the wave height where the water depth is 3 m?

3.2 Waves with a deep-water height $H_0 = 1.2$ m and period $T = 5$ s obliquely approach a shoreline with straight, parallel seabed contours. At a water depth $d = 6$ m, the wave height is $H = 0.9$ m. What is the shoaling coefficient for that depth? Estimate the deep-water wave direction α_0 (angle between wave crests and seabed contours).

3.3 A breakwater protecting a marina is sketched in the figure below. The incident wave train has a height $H = 1.5$ m and period $T = 3.6$ s. For the conditions indicated in the figure, what is the wave height at point A? What is the distance from the breakwater tip to a point P such that the wave height at P is H = 0.23 m? [*Use either Reference Solutions A3 or A4 in Appendix A.*]

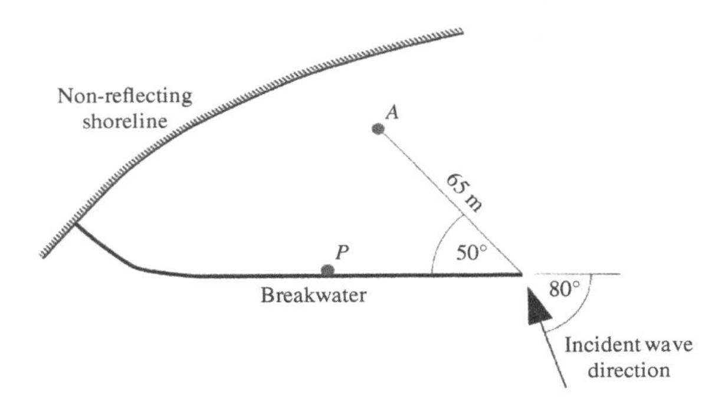

3.4 An incident wave train with height $H = 1.5$ m and period $T = 6.0$ s in water of uniform depth $d = 5.0$ m undergoes normal reflection at a vertical wall. Assuming perfect reflection, what is the shortest distance between the wall and a node of the resulting standing wave train? What are the maximum particle velocity and acceleration at the seabed at this location? What is the maximum particle velocity at the seabed at a distance of 5 m in front of the wall, and at what time does this occur (in seconds), given that time $t = 0$ when a wave crest is at the wall?

3.5 A harbor is approximated as an open-ended rectangular basin with length 800 m, width 100 m, and a uniform water depth of 15 m. What are the periods of first-mode oscillations in the longitudinal (800 m) direction and second-mode oscillations in the transverse (100 m) direction? For first-mode longitudinal oscillations with an incident wave height of 0.1 m, what are the orbital diameters of horizontal particle motions at the seabed, at the harbor entrance and at the closed end of the harbor?

3.6 ■ Waves in a wave flume are partially reflected by a beach, resulting in a combined incident and partially reflected wave field. Recognizing that the combined wave height varies with x along the flume, the incident wave height H_i and the reflection coefficient K_r are to be estimated by measurements of the maximum combined height, denoted H_{max}, and the minimum combined height, denoted H_{min}, along the flume. By considering an expression for the combined free surface elevation, develop expressions for H_{max} and H_{min} in terms of H_i and K_r. Given that H_{max} and H_{min} occur at $kx = 0, \pi, \ldots$ and at $kx = \pi/2, 3\pi/2, \ldots$, respectively, estimate H_i and K_r when $H_{max} = 0.16$ m and $H_{min} = 0.12$ m.

3.7 Waves obliquely approach a beach with straight parallel bottom contours and with a uniform slope $m = 1/5$. Waves are observed to break with a period $T = 8$ s at a location where the depth $d = 3$ m, with the breaking wave crests then making an angle $\alpha = 20°$ with the seabed contours. Using the relationship $H_b/d_b = 0.78$, estimate the breaking wave height, the deep-water wave height, and the form of the breaking waves.

3.8 Waves with a deep-water height $H_o = 1.5$ m, period $T = 7$ s, and deep-water direction $\alpha_o = 20°$ approach a beach with straight, parallel seabed contours and a uniform slope $m = 1/5$. Based on the breaking wave condition $H_b/d_b = 0.78$, estimate the depth at wave breaking, the breaking wave height, and the form of the breaking waves. [*You are advised to develop a table in which a series of decreasing values of kd are specified, and corresponding values of H/d are determined. The kd value at which the waves break may thereby be estimated.*]

4

Random Waves

4.1 Introduction

Individual waves vary from one to the next, so that there is a need to account for this variability in describing wave conditions during a particular sea state or storm that lasts for a few hours. Distinct from this short-term variability of individual waves, the intensity and frequency of storms vary during a year and from one year to the next, so that there is a need to account also for this storm variability in order to develop sufficiently extreme conditions for coastal engineering design. Thus, this chapter distinguishes between the short-term variability of individual waves within a few hours, considered in Sections 4.2 and 4.3, and the long-term variability of extreme storms over a number of years, considered in Sections 4.4–4.6. Finally, distinct from more extreme storms, the variability of different sea states in the medium term, represented by average wave conditions over a year, is also of interest with respect to some applications and is considered in Section 4.7.

With respect to short-term variability, Figure 4.1 highlights the variability of individual waves by comparing a photograph of a regular, periodic wave train in Figure 4.1a, with a photograph of an irregular or random water surface during a particular sea state in Figure 4.1b.

(a) (b)

Figure 4.1 Comparison of regular and random waves. (a) Regular, long-crested wave field, (b) random, short-crested wave field.

An Introduction to Coastal Engineering, First Edition. Michael Isaacson.

Companion website: www.wiley.com/go/coastalengineering

Thus, Figure 4.1a shows waves that may be approximated as a regular, long-crested wave train with a constant wave length, as was considered in Chapter 2. In comparison to this, Figure 4.1b shows a random or irregular wave field, indicating waves with individual wave heights and wave lengths that vary between successive waves and that are short-crested with no easily identified wave direction. These waves are also referred to as *multidirectional waves*, since the short-crestedness is associated with component wave trains with different directions. In this context, Sections 4.2 and 4.3 consider the treatment of a wave record with heights and lengths that vary from wave to wave, as well as waves that are short-crested.

The short-term variability of individual waves is commonly described by considering the probabilistic properties of individual wave heights and the spectral properties of waves that describes their frequency content. These are considered in turn below.

4.2 Probability Distribution of Wave Heights

Consider a wave record showing the variation of the water surface elevation at a given location as a function of time, as sketched in Figure 4.2. Individual wave heights and periods vary from one wave to the next. Thus, as shown in the figure, an individual wave height may be defined as the vertical distance between each wave trough and the subsequent wave crest, with two successive wave heights H_1 and H_2 indicated in the figure. An individual wave period may be defined as the duration between successive up-crossings, with two successive wave periods T_1 and T_2 indicated in the figure.

The variability of individual wave heights is commonly described by the probability distribution of heights. As background, a probability distribution is described as follows. The cumulative probability $P(x)$ of a continuous random variable x is the probability that any general value of x is less than the argument. Related to this, the exceedance probability $Q(x)$ is defined as $Q(x) = 1 - P(x)$ and is the probability that any general value of x is greater than or equal to the argument. Also, the probability density $p(x)$

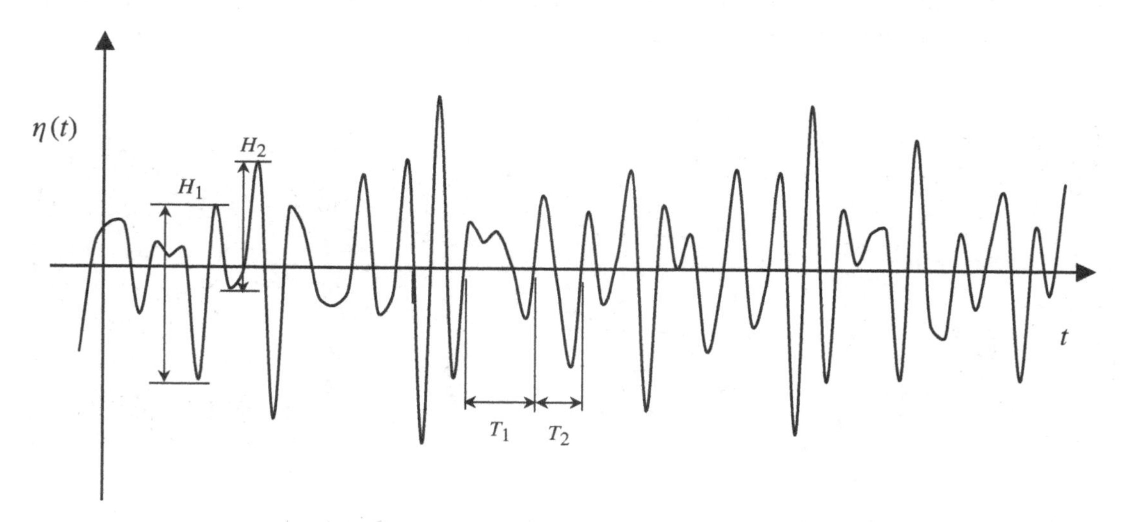

Figure 4.2 Sample of a random wave record.

is defined as $p(x) = \mathrm{d}P(x)/\mathrm{d}x$ so that $p(x)\mathrm{d}x$ is the probability that any general value of x lies between x and $x + \mathrm{d}x$.

With this background, the probability distribution of wave heights H is usually described by the Rayleigh distribution. This provides the cumulative probability in terms of a representative wave height H_r as follows:

$$P(H) = \begin{cases} 1 - \exp\left(-\dfrac{H^2}{H_\mathrm{r}^2}\right) & \text{for H} > 0 \\ 0 & \text{otherwise} \end{cases}$$

where the representative height H_r is the *root-mean-square wave height* that is defined as:

$$H_\mathrm{r} = \sqrt{\frac{1}{N}\sum_{i=1}^{N} H_i^2}$$

where N is the number of waves in a record and H_i are the individual wave heights.

Alternative representative wave heights may be used instead of H_r to scale the distribution. The most common one that is used is termed the *significant wave height*, H_s, which is defined as the average height of the highest one third of individual waves in the record. Based on the Rayleigh distribution, it may be shown that H_s and H_r are related by $H_\mathrm{s} = 1.416\, H_\mathrm{r}$ so that the Rayleigh distribution may be expressed in terms of H_s instead. Thus, given that $H \geq 0$, the cumulative probability $P(H)$, the exceedance probability $Q(H)$ and the probability density $p(H)$ are given in terms of H_s, respectively, by:

$$P(H) = 1 - \exp\left(-\frac{2H^2}{H_\mathrm{s}^2}\right)$$

$$Q(H) = \exp\left(-\frac{2H^2}{H_\mathrm{s}^2}\right)$$

$$p(H) = \frac{4H}{H_\mathrm{s}^2}\exp\left(-\frac{2H^2}{H_\mathrm{s}^2}\right)$$

Figure 4.3 shows plots of $P(H)$, $Q(H)$, and $p(H)$ based on the Rayleigh distribution.

Figure 4.3a shows that $P(H)$ increases from zero to one as H increases above zero, consistent with the limiting cases of zero probability for $H < 0$, and a probability of one for H less than a large value. Correspondingly, Figure 4.3b shows that $Q(H)$ reduces from one to zero as H increases above zero. Figure 4.3c shows the probability density plot, which corresponds to the slope of the cumulative probability plot, such that the area under the $p(H)$ curve is one.

A set of other representative wave heights may also be defined, with each one related to the others on the basis of the Rayleigh distribution. The relative magnitudes of these are indicated in Figure 4.3. Beyond H_r and H_s, these include the following, with their definitions and their values in terms of H_r and H_s given.

- mean or average wave height, $\overline{H}$: average of wave heights; $\overline{H} = 0.886\, H_\mathrm{r} = 0.626\, H_\mathrm{s}$
- median wave height, $\overline{\overline{H}}$: height at which $Q(H) = 0.5$; $\overline{\overline{H}} = 0.833\, H_\mathrm{r} = 0.588\, H_\mathrm{s}$
- mode of H, μ_H: height at which $p(H)$ is a maximum; $\mu_\mathrm{H} = 0.707\, H_\mathrm{r} = 0.50\, H_\mathrm{s}$
- standard deviation of η, σ_η: $\sigma_\eta = 0.354\, H_\mathrm{r} = 0.25\, H_\mathrm{s}$

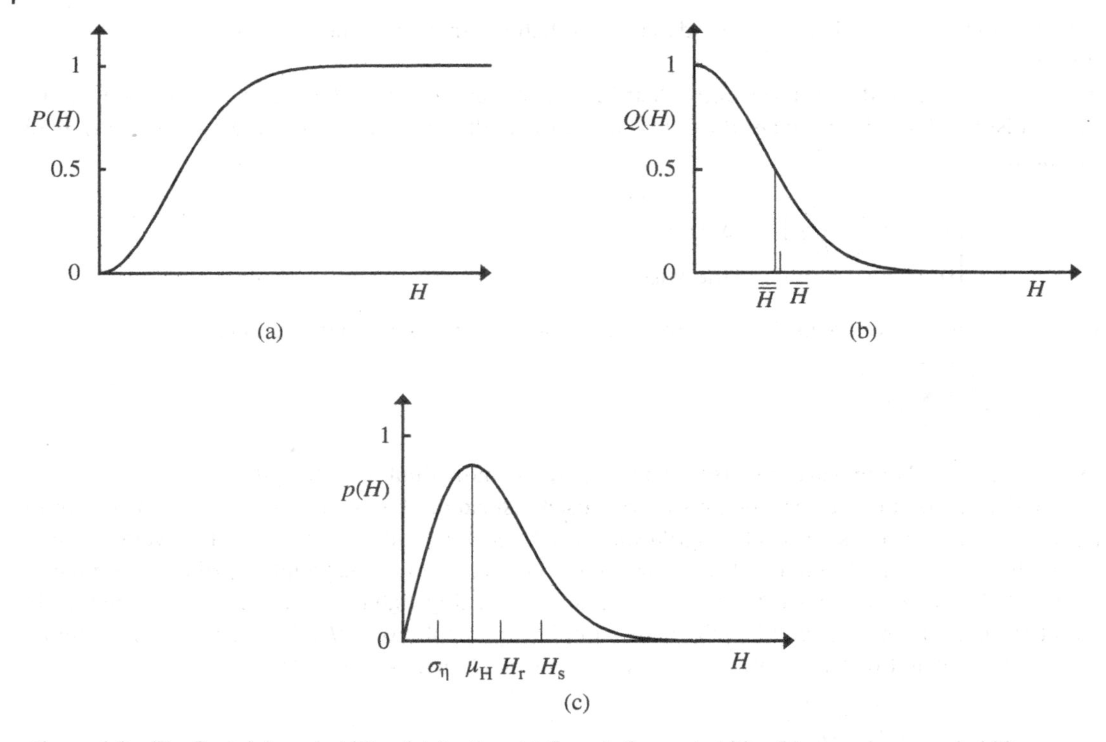

Figure 4.3 The Rayleigh probability distribution. (a) Cumulative probability, (b) exceedance probability, (c) probability density.

As indicated, if any one representative height is known, all the others may be determined.

The standard deviation of η, σ_η, is associated with η itself taken to follow a normal or Gaussian distribution with a zero mean. That is, the probability density of η is given by:

$$p(\eta) = \frac{1}{\sigma_\eta \sqrt{2\pi}} \exp\left(-\frac{\eta^2}{2\sigma_\eta^2}\right)$$

It is of particular interest to determine the most probable largest individual wave height H_m in a specified duration τ, since this is often used in applications. H_m depends on the number of waves N under consideration (the larger the sample, the larger may be the maximum wave height), which in turn depends on the duration τ being considered. H_m may be estimated from:

$$\frac{H_m}{H_s} = \sqrt{\frac{1}{2}\ln(N)} = \sqrt{\frac{1}{2}\ln(\tau/T)}$$

where $N = \tau/T$ is the number of waves, and T is a representative wave period. This ratio varies gradually with duration and may be approximated as: $H_m/H_s \simeq 1.8\text{–}2.0$.

Example 4.1 ***Probability Distribution of Wave Heights***
In a certain sea state, 15% of the waves have heights exceeding 1.2 m. What are the significant wave height and the mean wave height? What percentage of the waves has heights between 1.2 and 1.4 m? If $T = 4$ s, what is the largest wave height expected over a duration of 4 hr?

Solution
Specified parameters

$H_1 = 1.2\,\text{m}$
$H_2 = 1.4\,\text{m}$
$Q(H_1) = 0.15$
$T = 4.0\,\text{s}$
$\tau = 4.0\,\text{hr}$

Significant wave height and mean wave height

The formula for Q is $Q = \exp[-2(H/H_s)^2]$. This may be used to obtain H_s in terms of H_1 and Q:

$$H_s = H_1\sqrt{-2/\ln(Q)} = 1.23\,\text{m}$$

$$\overline{H} = 0.626\,H_s = 0.77\,\text{m}$$

Proportion of waves within a range of heights

$Q(H_1) = \exp[-2(H_1/H_s)^2] = 0.150$
$Q(H_2) = \exp[-2(H_2/H_s)^2] = 0.076$
Probability that $1.2\,\text{m} < H < 1.4\,\text{m} = 0.150 - 0.076 = 7.4\%$

Largest wave height

From formula for H_m:

$$H_m = H_s\sqrt{\frac{1}{2}\ln(\tau/T)} = 2.5\,\text{m}$$

4.3 Wave Spectra

4.3.1 One-Dimensional Spectra

As with wave heights, wave periods also vary from one wave to the next. However, while the probability distribution of wave periods may be developed in an analogous manner, a spectral description of the waves is instead adopted.

A wave spectrum provides a representation of the mix of wave frequencies in a wave record. The following provides a general description of a wave spectrum, while a more formal development is available in a number of texts, including Sarpkaya and Isaacson (1981) and Dean and Dalrymple (1991). In developing a wave spectrum, or, specifically, the spectral density $S(f)$, an irregular wave record $\eta(t)$ is considered to be composed of the superposition of component sinusoidal wave trains with different frequencies, amplitudes, and phases. It is well known that a periodic signal may be represented by a Fourier series, with components at multiples of a fundamental frequency. In this context, a random signal may be considered to be periodic, but with an infinite period so that it contains a continuous range of frequencies rather than discrete harmonics. Using such a representation, the spectral density $S(f)$ is defined such that $S(f)\,\mathrm{d}f$ is related to an equivalent Fourier amplitude squared at a frequency f. This corresponds to the contribution to the variance σ_η^2 of a record due to frequencies between f and $f + \mathrm{d}f$. The variance itself, indicative of the overall wave energy, is made up of contributions over the entire frequency range and thus corresponds to the area under the spectral density curve:

$$\sigma_\eta^2 = \int_0^\infty S(f)\,\mathrm{d}f$$

S is sometimes described in terms of the angular frequency ω rather than f, with $S(\omega) = S(f)/2\pi$.

As with wave heights, alternative representative wave periods may be developed. The two representative wave periods that are most commonly used are:

- *zero-crossing period,* T_z: average of all individual wave periods in a record, as based on successive up-crossings (see Figure 4.2)
- *peak period,* T_p: period at the peak of a wave spectrum

Also commonly used, the *peak frequency*, f_o, is the frequency at which the spectral peak occurs and is the reciprocal of the peak period, $f_o = 1/T_p$.

The relationship between T_z and T_p depends on the corresponding wave spectrum. For the Pierson–Moskowitz spectrum that is commonly used, this is $T_z = 0.71T_p$.

Various parametric formulations of the shape of the spectrum have been proposed. Initially, wave spectra were most often expressed in terms of wind speeds associated with wave generation. However, for the present, the shape of the spectrum is described in terms of the corresponding significant wave height and peak period.

One of the simplest and most common representations of the spectrum is referred to as the *Pierson–Moskowitz spectrum* that was proposed in 1964. While this spectrum was initially developed in terms of wind speed, it may be re-expressed in terms of significant wave height and peak period. The version expressed in this way is fully equivalent to a spectrum proposed in 1959 by Bretschneider. The corresponding spectrum, referred to in this text as the Pierson–Moskowitz spectrum, is given as:

$$S(f) = \frac{5H_s^2}{16f_o}\left(\frac{f}{f_o}\right)^{-5} \exp\left[-\frac{5}{4}\left(\frac{f}{f_o}\right)^{-4}\right]$$

A plot of this spectrum in dimensionless form is given in Figure 4.4.

4.3.2 Transformation of Wave Spectra

Recognizing that the concept of a spectrum applies to any time-varying signal, a spectral analysis may be applied to a linear system that relates an output signal to an input signal, whereby the spectrum of the output signal may be developed in terms of the spectrum of the input signal. In coastal engineering

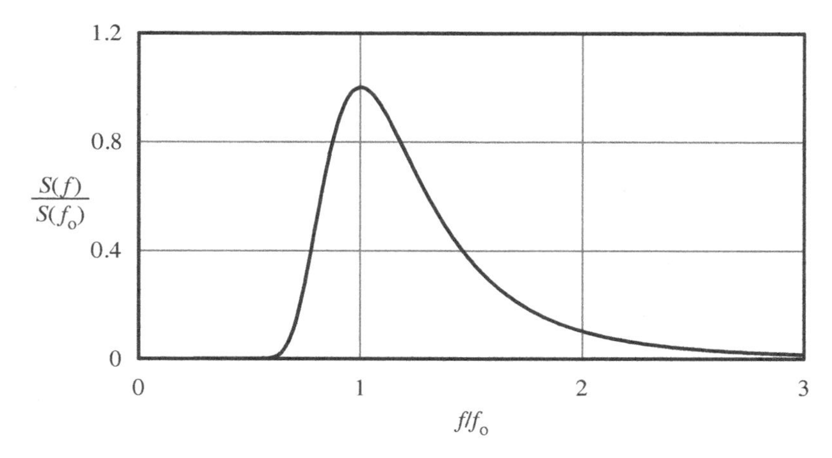

Figure 4.4 Plot of the Pierson–Moskowitz spectrum in dimensionless form.

applications, the incident wave field may be considered as an input signal, while examples of an output signal include the surface elevation of a transformed wave field (e.g. resulting from shoaling, refraction, and/or diffraction), a force on a structure, and a motion of a structure (e.g. the roll motions of a marine vessel). In such cases, the spectrum of an output variable x, denoted, $S_x(f)$, is related to the input wave spectrum $S_\eta(f)$ by:

$$S_x(f) = T_F(f)\, S_\eta(f)$$

where $T_F(f)$ is a *transfer function* that may be obtained from an analysis of regular wave behavior at a series of wave frequencies. Specifically, $T_F(f)$ is defined as $T_F(f) = (A_o/A_i)^2$, where A_o is the amplitude of the output parameter and A_i is the amplitude of the input parameter (i.e. incident wave elevation) at the same frequency. The multiplication described in the above equation is carried out over all frequencies in the manner illustrated in Figure 4.5.

The figure indicates how the multiplication is carried out at every frequency so that, for example, the peaks in the $S_\eta(f)$ and $T_F(f)$ curves carry over to the $S_x(f)$ curve to differing extents. The variance σ_x^2 of the output signal x corresponds to the area under the corresponding spectral density curve $S_x(f)$:

$$\sigma_x^2 = \int_0^\infty S_x(f)\, df$$

This leads to the standard deviation of x, σ_x, which may be considered to be a representative magnitude of the signal x. The standard deviation σ_x can be used in turn to estimate x_m, the maximum of x occurring over N waves or oscillations, in an analogous way to that for wave heights. Thus:

$$\frac{x_m}{\sigma_x} = \sqrt{2\ln(N)}$$

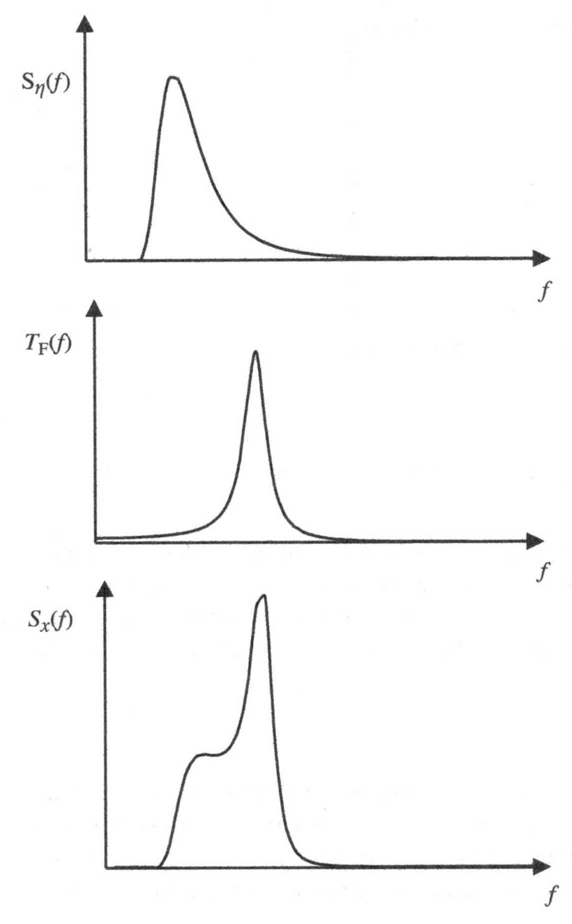

Figure 4.5 Transfer function $T_F(f)$ relating an output spectrum $S_x(f)$ to an input spectrum $S_\eta(f)$.

Example 4.2 ***Transformation of Wave Spectra***

Design wave conditions outside a marina are represented by a Pierson–Moskowitz spectrum with $H_s = 1.5$ m and $T_p = 4$ s. For a given wave direction, the diffraction coefficient K_d relating the regular wave height at a location A within the marina to the incident wave height is given as a function of wave frequency f by $K_d = 0.2/(f + 0.2)$, where f is in Hz (cycles per second). The incident waves correspond to an input signal, the square of the diffraction coefficient corresponds to the transfer function (since a spectrum is related to wave heights squared), and the waves at A correspond to the output signal. Superpose plots of the incident wave spectrum, the corresponding transfer function, and the wave spectrum at A. From the latter, estimate the significant wave height at A.

Such an approach may be extended to developing a wave field $\eta(x, y, t)$, corresponding to a prescribed directional spectrum $S(\omega, \theta)$. Thus, $\eta(x, y, t)$ is now expressed as the sum of component wave trains as:

$$\eta(x, y, t) = \sum_j \sum_k a_{jk} \cos(k_j x \cos\theta_k + k_j y \sin\theta_k - \omega_j t - \varepsilon_{jk})$$

where a_{jk} is the amplitude of a component wave train with frequency ω_j and direction θ_k, ω_j is obtained as $\omega_j = j\Delta\omega$, $\Delta\omega$ is a chosen small frequency interval, θ_k is obtained as $\theta_k = k\Delta\theta$, $\Delta\theta$ is a chosen small direction interval, k_j is the component wave number corresponding to ω_j, and ε_{jk} is random and evenly distributed between $-\pi$ and π. The amplitudes a_{jk} are obtained from the prescribed directional spectrum as:

$$a_{jk} = \sqrt{2S(\omega_j, \theta_k)\Delta\omega\Delta\theta}$$

←■

4.4 Long-Term Variability of Storms

Having examined the variability of individual waves over a few hours, attention is now directed to the variability of successive storms or sea states over a number of years in order to determine a design wave height representative of an extreme storm condition. It is now the probability distribution of a series of sea states or storms, each one represented by a single parameter H, that is of interest. H now denotes a representative wave height of a sea state or storm – most commonly the maximum value of H_s that occurs within a storm. This probability distribution needs to be estimated and then used to obtain a design value of H corresponding to a required remote probability of exceedance. The latter is usually expressed in terms of a specified *return period* T_R, which is the average duration between successive exceedances of the given magnitude of the event occurring. Related to this, reference is also made to the *annual exceedance probability* (AEP), which is the reciprocal of the return period. For example, a wave height with a 50-yr return period is exceeded on average once every 50 years and has an AEP value of 2%. Additional terminology that is sometimes used may refer to this wave height as a "50-yr wave height" or a "1-in-50-yr wave height."

The above procedure, termed an extreme value analysis (EVA), is developed in Section 4.5 with respect to wave data. An EVA may be also applied to other extreme natural phenomena, notably wind speeds and flood levels.

4.5 Extreme Value Analysis

4.5.1 Overview

The most common approach is to identify the most extreme storms over a number of years, each one defined by a representative height H, use this information to estimate the long-term probability distribution of H, and then apply this probability distribution to estimate the wave height corresponding to a more remote probability, usually taken to correspond to a specified return period. This process is equivalent to what is referred to as the *peak-over-threshold method*, in that it relies on extracting peak values exceeding some threshold over a particular duration. This is in contrast to a different method that is based on annual maxima.

In summary, the following steps are taken:

1. ***Exceedance probabilities***. A table is developed of the most severe H values ranked in order of decreasing severity, along with corresponding exceedance probabilities Q.
2. ***Distribution selection and fit***. A selected extreme probability distribution is fitted to the H–Q data, suitably scaled by transforming them to corresponding x and y values such that the selected distribution is fitted by a straight line.
3. ***Return period and AEP***. A T_R–Q relationship is established so that, for any prescribed T_R or AEP, the required Q is known. The fitted line can thereby be used to obtain the value of H corresponding to the prescribed T_R or AEP.
4. ***Encounter probability***. In some cases, the encounter probability E over the design life L of the project is required and may be obtained in terms of L and T_R.

A description of each of these steps is as follows.

4.5.2 Exceedance Probabilities

The exceedance probability $Q(H) = 1 - P(H)$ is the probability that a general wave height is greater than the argument H. The N most severe storms over a specified duration τ are identified and the corresponding H values are ranked, with $n = 1$ corresponding to the largest H, $n = 2$ corresponding to the second largest H, and so on. The selection of N is a matter of judgment and corresponds to the number of data points representing the tail of the distribution that is used as a basis for extrapolation. The exceedance probability $Q(H)$ for each of these H values is then obtained from:

$$Q(H) = \frac{n}{M}$$

where M can be any large number corresponding to the total number of storms above some threshold over the duration τ. That is, only the N most extreme storms of the total number M are used in the plotting procedure.

4.5.3 Distribution Selection and Fit

The above step provides a table of ranked H values and corresponding Q values. The next step is to fit a curve to the H–$Q(H)$ data. However, a linear scale would make extrapolation difficult and inaccurate. Because of this, a different scale based on an extreme probability distribution is chosen such that the selected distribution plots as a straight line.

Extreme distributions that have been used, along with the x and y values in which these plot as straight lines, are provided in Table 4.1. In the table, ε is a free parameter selected to give the best fit to data and $\mathrm{erf}(y)$ is the error function:

$$\mathrm{erf}(y) = \frac{2}{\sqrt{\pi}} \int_0^y \exp(-t^2)\, \mathrm{d}t$$

Table 4.1 Selected extreme value probability distributions.

Distribution	Abscissa x	Ordinate y
Lognormal	$\ln(H)$	$1 - Q(H) = \text{erf}(y)/2$
Gumbel	H	$-\ln[-\ln(1 - Q(H))]$
Frétchet	$\ln(H)$	$-\ln[-\ln(1 - Q(H))]$
Weibull (lower bound)	$\ln(H - \varepsilon)$	$\ln[-\ln(Q(H))]$
Weibull (upper bound)	$-\ln(\varepsilon - H)$	$-\ln[-\ln(1 - Q(H))]$

The Gumbel, Frétchet and Weibull distributions are also referred to as Type I, II, and III distributions, respectively, of a family of generalized extreme value distributions developed through extreme value theory. Of these distributions, the Gumbel distribution is the most commonly used. Once the distribution has been selected, the table of ranked H values and corresponding Q values is extended to include additional columns with corresponding x and y values based on the formulae given in the above table. The pairs of x and y values may thereby be plotted and a best-fit line obtained.

Figure 4.8 shows an example plot arising from an EVA. The figure shows data points (filled circles), the fitted line, and the 100-yr wave height (unfilled circle). This particular example is based on the Weibull distribution to highlight the distinction between x and H along the x axis, and it also shows the distinction between Q and y along the y axis. The figure also indicates return periods, since these are related to Q values in a manner that is given below.

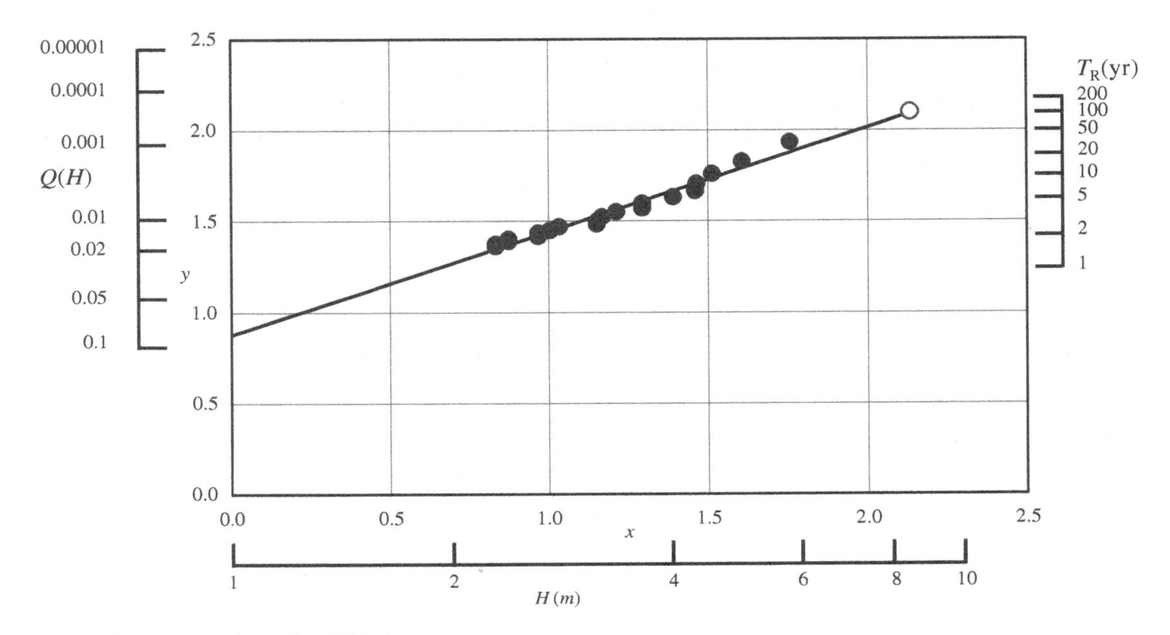

Figure 4.8 Illustration of an EVA plot.

It is noted that the data itself may exhibit scatter and that a straight line may not suitably fit the data over the entire range. In the latter case, it is the fit at the upper end of the data that is most relevant with respect to undertaking the extrapolation.

4.5.4 Return Period and Annual Exceedance Probability

As mentioned earlier, the return period is defined as the average duration between successive design events being exceeded. Related to this, the AEP is defined as the reciprocal of the return period. The straight-line fit indicated above needs to be extrapolated so as to obtain a design wave height corresponding to a specified return period. In order to do so, it is necessary to relate the Q values to return periods. First, a time scale needs to be developed for the data. This is the *recording interval* r_i, which is the average duration between adjacent H values over the totality of storms. That is, r_i is given by $r_i = \tau/M$, where M is the total number of data values. (In fact, the denominator ought to be $M - 1$, but this is of no consequence since M is arbitrary and large.)

Once the recording interval has been obtained, the return period can be expressed in terms of $Q(H)$ as:

$$\frac{T_R}{r_i} = \frac{1}{Q}$$

Therefore, the value of Q corresponding to a specified value of T_R is known, so that the straight-line fit can be used to determine the corresponding x value and thereby the corresponding wave height H.

It is noted that, since r_i and Q are both inversely proportional to M for larger values of M, the selection of M does not affect the H–T_R relationship.

4.5.5 Encounter Probability

As discussed, the likelihood of an extreme event (e.g. design wave height) being exceeded is commonly expressed in terms of a return period. However, there is usually a need to assess the probability of such an event being exceeded during the *design life* L, which is the specified duration for which the project is designed to continue to function effectively. Thus, the *encounter probability* E is the probability that an event with a specified return period T_R is equaled or exceeded during the design life L. E is related to T_R and L by:

$$E = 1 - \left(\frac{1}{T_R/r_i}\right)^{L/r_i}$$

For the usual case when both L/r_i and T_R/r_i are large, this may be simplified to:

$$E = 1 - \exp(-L/T_R)$$

For example, the probabilities of a 75-yr return-period event occurring within a 75-yr duration and a 10-yr duration are 63% and 12%, respectively. When codes and standards do not explicitly refer to the basis for selecting a design event, the onus is on the designer to select a suitable return period or encounter probability.

Reference Solution A5 in Appendix A provides a spreadsheet solution based on the EVA described above.

■➔

Example 4.3 ***Extreme Value Analysis – Spreadsheet Example***
The 12 most extreme storms over the eight years 2017 to 2024 have maximum significant wave heights H as follows: 1.55, 2.00, 1.90, 2.12, 1.80, 2.20, 1.96, 2.08, 2.35, 1.70, 1.89, and 1.92 m. Estimate H corresponding to a return period $T_R = 25$ yr based on the entire data set, and, again, based on a suitable subset of the data.

Solution
Using Reference Solution A5, proceed as follows:

- Enter duration $\tau = 8$ yr and return period $T_R = 25$ yr
- Adjust the table template so as to correspond to 12 rows. Then, paste the ranked heights into column C
- The resulting plot, as shown below, provides the data points (filled circles), the fitted line (broken), denoted fit-A, and the 25-yr height (unfilled circle), denoted H(25)-A, which is equal to 2.8 m
- The plot indicates that the lowest three data points do not suitably fit the tail of the data and so are omitted
- The table template is then adjusted so as to correspond to nine rows, with the lowest three heights omitted
- The resulting plot now provides a modified fitted line (solid), denoted fit-B, and the modified 25-yr height (shaded circle), denoted H(25)-B, which is equal to 2.6 m

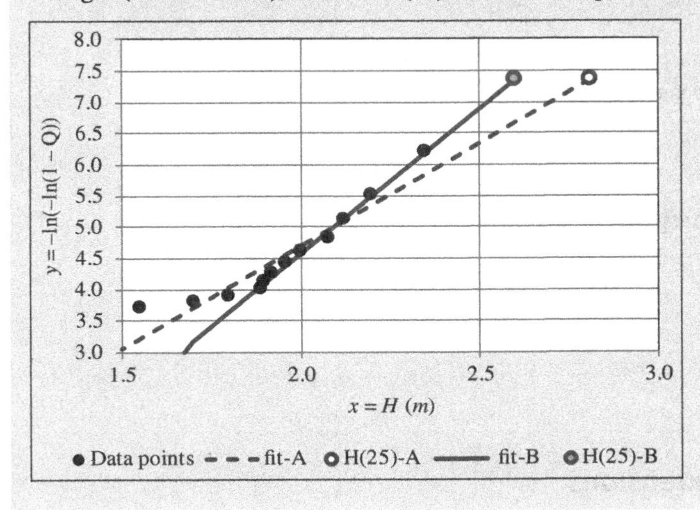

Example 4.4 ***Extreme Value Analysis – Algebraic Example***
The significant wave heights of the two most severe storms at a site over a 10-yr period are 1.8 and 2.2 m. Assuming that the significant wave heights follow a Gumbel distribution, estimate the significant wave height that has a return period of 50 yr.

Solution

Specified parameters

$H_1 = 2.2\,\text{m}$
$H_2 = 1.8\,\text{m}$
$\tau = 10\,\text{yr}$
$T_R = 50\,\text{yr}$

Develop table of (x, y) values

Select any large value of M, say $M = 100$. Therefore:
$r_i = \tau/M = 0.1\,\text{yr}$

Obtain corresponding Q, $x = H$ and $y = -\ln(-\ln(1-Q))$ values:

H (m)	Rank n	Q [$= n/M$]	x [$= H$]	y [$= -\ln(-\ln(1-Q))$]
2.2	1	0.01	2.2	4.600
1.8	2	0.02	1.8	3.902

Straight-line fit

A straight-line fit based on $y = mx + c$ is used to obtain m and c:

Slope $m = (y_2 - y_1)/(x_2 - x_1) = 1.746$
Intercept $c = y_2 - mx_2 = 0.760$

For $T_R = 50$ yr, obtain in turn the corresponding Q, y, x, and H values:

$Q = r_i/T_R = 0.002$
$y = -\ln(-\ln(1 - Q)) = 6.214$
$x = (y - c)/m = 3.124\,\text{m}$
$H = x = 3.1\,\text{m}$

■➔

4.6 EVA Alternatives and Extensions

The preceding description of an EVA is not unique, and various alternatives and extensions to it may be considered.

4.6.1 Annual Maxima

An alternative approach, which is used in hydrology with respect to flood levels that typically peak once a year, is to develop the series of events as annual maxima. This implies that the recording interval is one year, so that the most severe event each year is used in the EVA. This is entirely appropriate with respect to flood levels but not so with respect to coastal storms, since there may be more than one relatively severe storm in a given year.

4.6.2 Lower Return Periods

In some instances, it may be necessary to develop wave conditions corresponding to lower return periods of one year or less. In such cases, the same EVA methodology may continue to be used, except that, first, the data on which the analysis is based would need to entail a recording interval that is lower than the return period of interest; and second, an interpolation of any fit to the data that envelopes the desired return period should be undertaken in lieu of an extrapolation based on a fit to the tail of the data. However, for more commonly occurring wave conditions, the approaches described in Section 4.7 are preferred.

4.6.3 Seasonal Conditions

There may be an interest in examining wave conditions appropriate to a portion of each year, for example with respect to the open-water season at locations that are ice-covered during the winter, or a boating season that extends over the summer months only. Thus, a 50-yr wave height, say, with respect to the summer season will be lower than the corresponding year-round wave height. In such cases, the same EVA methodology may be used, except that the data used to develop the ranked list of wave heights needs to correspond to the relevant period for each year covered by the dataset.

4.6.4 Confidence Bands

It is noted that the method of extrapolating wave data to obtain a design wave height assumes that the fitted curve is valid, whereas the fit itself has some uncertainty. In order to explore this aspect, confidence bands may be plotted around the best-fit line used to approximate the data. These bands describe the closeness of fit of the data points to the fitted distribution, expressed as confidence limits about either side of the fitted line. For example, a 90% confidence band may be plotted as a pair of lines above and below the best-fit line such that, for any given $Q(H)$, the lines indicate the range of H values that have a 90% probability of occurring.

4.7 Annual Wave Conditions

4.7.1 Wave Scatter Diagram

Distinct from a consideration of extreme waves, some aspects of a project may rely on more commonly occurring wave conditions. For example, such information may be required with respect to downtime estimates of certain operations, fatigue calculations, and criteria of acceptable wave conditions within marinas. A description of the wave climate throughout the year is usually presented in tabular form showing average annual percentage occurrences of different combinations of significant wave height H_s, peak period T_p, and wave direction θ. Such information may be conveyed in different ways. Specifically, a *directional wave scatter diagram* provides different $H_s - \theta$ combinations, with rows corresponding to different H_s ranges (with the predominant T_p for each H_s value also shown), columns corresponding to different wave directions, typically based on 16 compass points (north, north-northeast, ...), and each cell containing the percentage occurrence in a year of each height–direction combination. If relevant, such information can be provided for different months or portions of a year.

When wave direction is not considered, a *wave scatter diagram*, which is illustrated in Figure 4.9, provides different $H_s - T_p$ combinations, with rows corresponding to different H_s ranges, columns corresponding to different T_p ranges and each cell containing the percentage occurrence of each $H_s - T_p$ combination. The figure also shows percentage occurrence summations for each wave height range and for each wave period range, with either of these sums totaling 100%. In utilizing such information, each cell's value is attributed to the central values of the corresponding H_s or T_p ranges.

The above information may be developed from sea state measurements over a number of years, or by a numerical hindcast model that develops hourly wave conditions continually over several years on the basis of wind records. (Wave hindcasting is described in Chapter 6.)

4.7.2 Long-Term Distribution of Individual Wave Heights

It may be of interest to estimate the long-term distribution of individual wave heights H. This may be developed by suitably combining the short-term distribution of individual wave heights H for a given significant wave height H_s, described by the Rayleigh distribution, with the long-term distribution of significant wave heights H_s described by a scatter diagram. This needs to take account of a weighting of the number of waves for each wave scatter diagram entry. Battjes (1970) has thereby developed an expression for the exceedance probability of individual wave heights H, denoted $Q_L(H)$, over the long term in terms of suitable integrations over H_s and wave period. When applied to scatter diagram data,

H_s (m)	T_p (s)											SUM:
	1.5–2.0	2.0–2.5	2.5–3.0	3.0–3.5	3.5–4.0	4.0–4.5	4.5–5.0	5.0–5.5	5.5–6.0	6.0–6.5	6.5–7.0	
2.2–2.4	0	0	0	0	0	0	0	0	0	0	0	0.00
2.0–2.2	0	0	0	0	0	0	0	0	0	0.04	0	0.04
1.8–2.0	0	0	0	0	0	0	0	0	0.09	0.00	0.02	0.11
1.6–1.8	0	0	0	0	0	0	0	0.02	0.05	0.08	0.03	0.18
1.4–1.6	0	0	0	0	0	0.00	0.01	0.04	0.18	0.05	0.01	0.29
1.2–1.4	0	0	0	0	0	0.01	0.05	0.34	0.40	0.08	0.01	0.89
1.0–1.2	0	0	0	0	0	0.10	0.53	0.61	0.45	0.03	0	1.72
0.8–1.0	0	0	0	0	0.20	1.56	1.95	0.95	0.51	0	0	5.17
0.6–0.8	0	0	0	0.14	1.70	5.64	4.16	2.01	0.88	0.16	0	14.69
0.4–0.6	0	0	0.04	1.66	7.56	10.84	5.92	2.28	1.31	0.44	0	30.05
0.2–0.4	0	0.04	3.42	6.90	11.88	12.25	5.90	1.08	0.41	0	0	41.88
0.0–0.2	0	0.19	0.87	1.40	1.55	0.74	0.19	0.04	0	0	0	4.98
SUM:	0.00	0.23	4.33	10.10	22.89	31.14	18.71	7.37	4.28	0.88	0.07	100.00

Figure 4.9 Illustration of a wave scatter diagram.

the expression for $Q_L(H)$ becomes:

$$Q_L(H) = \frac{\sum_i \sum_j \exp\left[-2H/H_{si}^2\right] T_j^{-1} w_{ij}}{\sum_i \sum_j T_j^{-1} w_{ij}}$$

where w_{ij} are the scatter diagram entries for the i-th H_s range and the j-th T_z range, H_{si} is the central H_s value of the i-th H_s range, T_j is the central T_z value of the j-th T_z range, and the summations are carried out over all values of i and j. It is noted that, in order for the numbers of waves for each scatter diagram cell to be accounted for correctly, the above expression relates to zero-crossing periods T_z and not peak periods T_p, so that a conversion from T_p to T_z, typically given by $T_z = 0.71T_p$, may need to be applied.

4.7.3 Application to Hours Per Year

In some coastal engineering applications, an estimate of the significant wave height that is not exceeded for a specified number of hours per year is needed. For example, this arises with respect to one form of available criteria relating to acceptable wave heights in a marina.

This kind of information may readily be developed from a given wave scatter diagram. First, the specified number of hours per year, denoted N_h, is used to determine a corresponding exceedance probability, based on $N_h = 8766\ Q(H_s)$, so that H_s corresponding to a given $Q(H_s)$ is needed. A scatter diagram is used to develop a table of $H_s - Q(H_s)$ values, a curve is then fitted to this data analogous to the EVA procedure described earlier, and an interpolation or extrapolation of this curve may then be undertaken to provide the H_s value corresponding to the given $Q(H_s)$.

4.7.4 Application to Fatigue Calculations

Numbers of individual waves with different periods and heights that occur per year have application to fatigue life calculations. As background, fatigue calculations are commonly based on a known S–N curve, which defines the number of cycles to failure, N, for different stress amplitudes S. This is applied in combination with Miner's rule that describes how different numbers of cycles for a series of different stress amplitudes contribute to fatigue damage. The latter refers to the percentage of the fatigue life that has been consumed over a certain duration. As one variant of this approach, the cumulative fatigue damage per year, denoted D_1, is given as:

$$D_1 = \sum_i \frac{n_i}{N_i}$$

where the summation is carried out over all possible stress amplitudes S_i, N_i is the number of cycles to failure corresponding to S_i, obtained from a given S–N curve, and n_i is the number of cycles occurring in a year at a stress amplitude S_i.

Given that stress amplitudes are proportional to wave heights with a known proportionality constant, the cumulative fatigue damage per year may be estimated from knowledge of the numbers of cycles of individual wave heights occurring during a year. This information cannot be developed from the long-term distribution of individual wave heights given above, since that does not provide sufficient information relating to numbers of cycles. Instead, a table of the number of cycles per year, N_k, corresponding to a series of individual wave heights H_k is required. This may be developed from a known scatter diagram as described below.

The following notation is used: the index i refers to the i-th significant wave height range, with a central value H_{si}, the index j refers to the j-th zero-crossing period range with a central value T_{zj}, and the index k refers to the k-th individual wave height range with a central value H_k. w_{ij} denotes the scatter diagram entry corresponding to $H_{si} - T_{zj}$, expressed as a percentage so that the w_{ij} values sum to 100%. Finally, Δ denotes the wave height interval, which is taken as constant.

The calculation proceeds as follows. First, since the number of cycles depends on the zero-crossing period T_z rather than the peak period T_p, a conversion of the scatter diagram in terms of T_z is undertaken. Then, each cell in the scatter diagram is used to develop the number of cycles per year n_{ij} corresponding to each $H_{si} - T_{zj}$ combination. The Rayleigh distribution is then used to determine the probability P_{ik} of an individual wave height H_k occurring, given a significant height H_{si}. Finally, the number of cycles per year, n_k, corresponding to each H_k value may be obtained in terms of n_{ij} and P_{ik}. The corresponding expressions are as follows:

$$n_{ij} = \frac{3600}{T_j} \frac{8766\, w_{ij}}{100}$$

$$P_{ik} = \Delta \frac{4H_k}{H_{si}^2} \exp\left(-\frac{2H_k^2}{H_{si}^2}\right)$$

$$n_k = \sum_i \left(P_{ik} \sum_j n_{ij}\right)$$

←■

Problems

4.1 A random wave train has an average wave height $\overline{H} = 0.7\,\text{m}$ and a period $T = 5\,\text{s}$. What is the largest individual wave height that is expected over a 1-hr duration? What is the probability that an individual wave height exceeds 1.6 m? The waves are perfectly reflected by a vertical seawall that extends a distance 1.6 m above the still water level. How often on average would you expect wave overtopping to occur? Express your answer in minutes between successive wave crests overtopping the wall on average.

4.2 The text's companion site allows for a download of the spreadsheet *Problem Data*. The tab "Problem 4.2" provides a sequence of 100 individual wave heights measured at a certain location. Compute the following parameters directly from the data: the mean wave height, the significant wave height, the root-mean-square wave height, and the largest wave height. What would be the estimated significant wave height and the estimated largest wave height that you would obtain from the mean wave height that you have computed? Show your results to three decimal places.

4.3 On the basis of the Rayleigh distribution, derive an expression for the root-mean-square wave height H_r in terms of the largest and second largest individual wave heights, denoted H_1 and H_2, respectively, that occur in a record of N individual waves. Hence, estimate the significant wave height H_s for the case $H_1 = 3.1$ m and $H_2 = 2.8$ m.

4.4 A sea state is described by a Pierson–Moskowitz spectrum with $H_s = 1.5$ m and $T_p = 4.0$ s. Plot the spectrum. What is the maximum value of the spectral density? What percentage of the wave energy density corresponds to frequencies within ±20% of the peak frequency?

4.5 ■ An incident wave train described by the Pierson–Moskowitz spectrum with $H_s = 1.2$ m and $f_o = 0.2$ Hz approaches a long, narrow inlet with a harbor located at its inner end A. Because of the inlet's resonance characteristics, the wave spectrum at A is related to the incident wave spectrum through an idealized transfer function $T_F(f)$ as shown in the figure below, with the waves at A amplified near $f = 0.25$ Hz but they are otherwise absent. What is the value of the incident wave spectrum (in units of m^2/Hz) at a frequency of 0.25 Hz? Without calculations, sketch the incident wave spectrum, the transfer function, and the wave spectrum at A. Estimate the standard deviation of the water surface elevation at A and the significant wave height at A. *[A multiplication at a single frequency only is required.]*

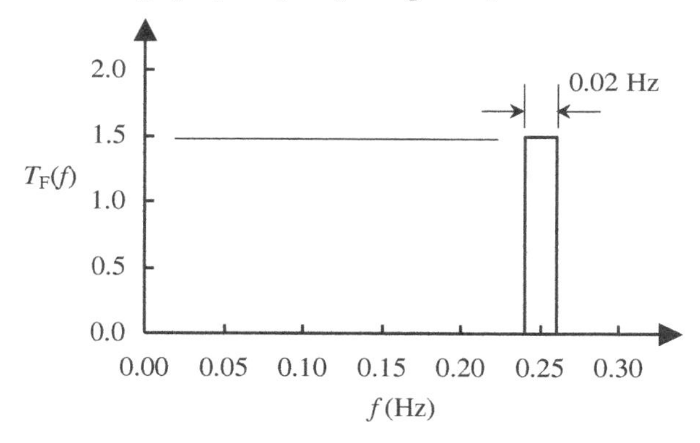

4.6 The maximum significant wave heights H of the five most extreme storms over a 10-yr period are as follows: 1.78, 1.68, 1.85, 2.12, and 1.94 m. On the basis of the Gumbel distribution, estimate H corresponding to a return period of 50 years. *[Use Reference Solution A5 in Appendix A to develop your response.]*

4.7 ■ The 10-yr and 25-yr design wave heights at a site are 1.8 and 2.1 m, respectively. Estimate the 100-yr design wave height by assuming in turn a Gumbel distribution, a Frétchet distribution, and a lower-bound Weibull distribution with $\varepsilon = 0$. Show your results to two decimal places.

4.8 ■ A short-crested sea state is described by a unidirectional wave spectrum combined with a cosine-squared directional spreading function that has a spreading index s. Develop an integral expression in terms of s and θ_o for the percentage of wave energy that is associated with wave directions within $\pm\,\theta_o$ of the mean wave direction. Use a spreadsheet to estimate this percentage when $s = 4$ and $\theta_o = 10°$.

4.9 ■ The text's companion site allows for a download of the spreadsheet *Problem Data*. The tab "Problems 4.9–4.10" contains a table corresponding to the wave scatter diagram shown in Figure 4.9. In the context of acceptable wave conditions within a marina, use this information to estimate the threshold wave heights that are exceeded no more than 15 and 76 hr per year in turn.

4.10 ■ The text's companion site allows for a download of the spreadsheet *Problem Data*. The tab "Problems 4.9–4.10" contains a table corresponding to the wave scatter diagram shown in Figure 4.9. In the context of fatigue calculations, use this information to develop a table of the number of cycles per year N_k associated with individual wave heights H_k corresponding to 0.1, 0.3, … , 3.5 m. Estimate the total number of individual waves per year and the number of individual waves per year with wave heights greater than 1.0 m.

5

Winds

5.1 Introduction

Coastal engineering practice requires an understanding of wind fields for several reasons. A knowledge of wind fields is needed in order to estimate wind-generated waves (which is often a central requirement of coastal engineering projects), storm surge (which refers to raised water levels during a storm, arising primarily from winds blowing over the sea surface), and wind-generated currents. As well, wind loads acting on boats and marine vessels result in increased loads on docks and other facilities, and need to be taken into account in a mooring system design. In this context, this chapter describes wind fields as required and utilized in coastal engineering.

5.2 Wind Data

The wind speed and wind direction at one location vary continually with time and with elevation above the sea or earth's surface and they also vary from one location to the next. Wind records are available for many wind stations. Typically, wind data is gathered in the form of a 1-min or 2-min average wind speed along with the corresponding wind direction, often at intervals of 10°, gathered every hour. The corresponding speed is taken to represent conditions over the entire hour and so is referred to as an hourly wind speed or a 1-hr wind speed. Because wind speed varies notably with elevation, most data is obtained at or normalized to an elevation of 10 m above the surface.

Such data gathered over an extended duration may be used to develop various kinds of summary information and to undertake different kinds of analyses. In various countries, hourly wind data for different stations may be downloaded from sites relating to government agencies, volunteer organizations, and/or companies. In the US, hourly wind data may be downloaded from the National Oceanic and Atmospheric Administration (NOAA) website ncei.noaa.gov and from a number of companies. In Canada, hourly wind data may be downloaded from Canada's Historical Climate Data website climate.weather.gc.ca and from the volunteer-run website weatherstats.ca.

Winds blowing over a fetch adjacent to a coastal location are needed to estimate resulting wave conditions. While these winds are usually taken to correspond to wind records at a wind station nearby, it is noted that wind speeds and directions may vary notably with location because of a prevailing weather

An Introduction to Coastal Engineering, First Edition. Michael Isaacson.

Companion website: www.wiley.com/go/coastalengineering

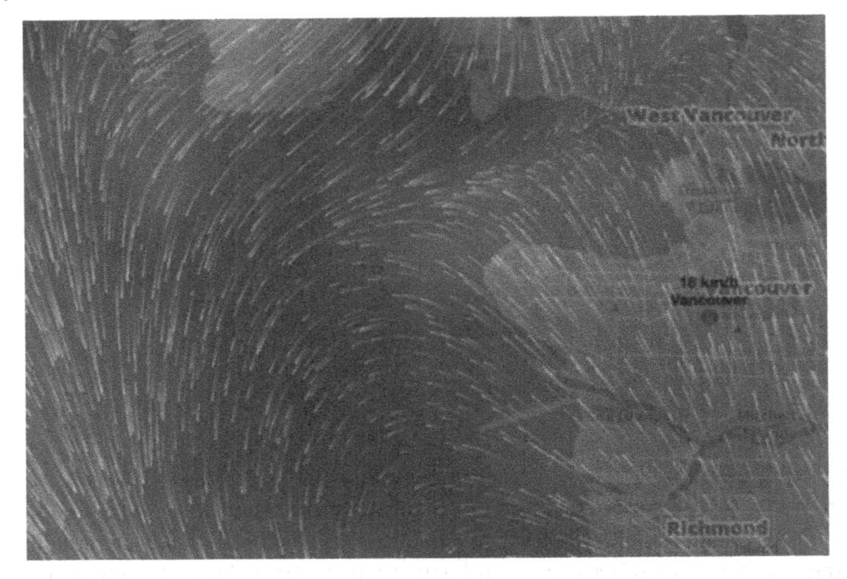

Figure 5.1 Illustration of wind direction variation near Vancouver, BC. *Source:* Reproduced with permission of Ventusky/InMeteo.

pattern and/or when the fetch is bounded by mountainous terrain. To illustrate the changes in wind direction that may occur, Figure 5.1 shows wind vectors at a particular day and time in the vicinity of Vancouver, BC, obtained from an app available at www.venusky.com. The figure shows notable changes in wind direction, including southeasterly winds at Richmond (lower right of the figure), easterly winds about 5 km west of Vancouver (center of the figure), and southerly winds about 15 km west of Vancouver (right of the figure).

In order to make allowances for changes in wind speed and direction, it may be possible to rely on measurements at more than one wind station. As well, information on wind fields available through different companies and apps may be used to provide indications of wind speed and direction changes that could be accounted for in an analysis.

5.3 Annual Wind Conditions

In addition to extreme wind speeds, information on commonly occurring wind conditions throughout the year may also be needed for design. Such information is obtained in the form of annual average wind statistics and may be presented in tabular or graphic form. This is usually provided in the form of a table with rows corresponding to different wind directions, typically based on 16 compass points (north, north–northeast, ...), columns corresponding to different wind speed ranges, and with each cell containing the percentage occurrence throughout a year of each speed-direction combination. (That is, when the percentage occurrence of calm conditions is included, the cell values should add up to 100%.) Note that wind direction refers to the direction from which the wind is blowing. If relevant, such information can also be provided for portions of a year.

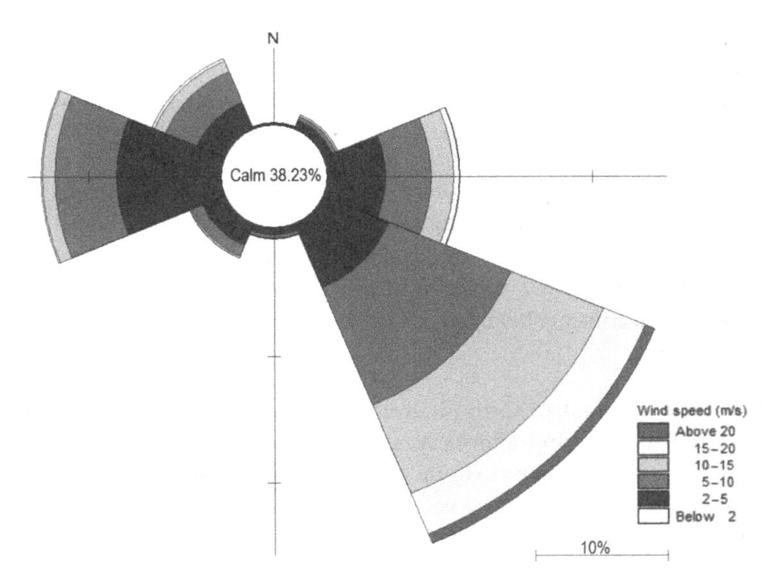

Figure 5.2 Illustration of a wind rose. *Source:* Reproduced with permission of CMO Consultants Ltd.

A common form of representation of such data is through a *wind rose*, as illustrated in Figure 5.2. This shows the percentage distributions of wind speeds and directions at a specified wind station for a typical year, or for each month of the year, or for some other time period. The development of such a representation from tabular wind data is straightforward.

5.4 Design Wind Speeds

A wind rose provides distributions of wind speed for an average year or month but does not provide extreme wind speeds that are needed in design. Therefore, there is a need to develop suitably defined extreme wind conditions from wind data that have been gathered over several years in the form of hourly wind speeds and directions.

In particular, design wind speeds corresponding to different directions and durations are needed in order to undertake a simplified hindcast analysis as described in Section 6.2. The analysis of wind station data to provide such information is now described.

Threshold and direction filters. Bearing in mind that an extreme value analysis (EVA) relies only on data for the highest speeds, downloaded hourly wind speed and direction data is first filtered so as to reduce the size of the data file by removing all data below a selected threshold value (e.g. 30 km/hr) and by removing all data except for winds from a specified range of directions. If wind data is provided in 10° direction increments, it may be prudent to allow for directional uncertainty by attributing winds from adjacent directions to the required direction. As an example, winds from 20°N to 40°N are retained in order to develop data valid for 30°N.

Wind duration. In a simplified wave hindcasting analysis, a wind speed that persists above a certain duration (e.g. 2 or 3 hr), referred to as a multi-hour wind speed, is required. Therefore, a wind data file may need to be processed to obtain, say, 3-hr wind speeds by considering, for each hour, the minimum wind speed over three consecutive hours.

Clustering effect. Each storm corresponds to a series of hourly wind speeds, whereas the storm needs to be represented by a single data value with respect to undertaking an EVA. Therefore, there is a need to avoid a clustering effect, in which two or more data points from the same storm are inadvertently used. Because of this, it is necessary to require a suitable time interval between successive data points that are used in the EVA, so as to assure that distinct storms are being considered and that only the largest wind speed within each storm represents that storm.

Extreme value analysis. An EVA is then undertaken in the manner described previously for wave conditions. The following approach is not unique and alternatives are possible. A set of N suitably defined wind speeds, denoted U_{n}, that represent the most extreme storms, are extracted from the data and ranked, with n ranging from 1 (highest speed) to N. The corresponding values of the exceedance probability Q_{n} are then obtained from $Q_{\mathrm{n}} = n/M$, where M is any large number representing the totality of storms over the duration τ of data measurements. The N pairs of values (U_{n}, Q_{n}) are fitted by a selected extreme probability distribution. A design wind speed corresponding to a specified (remote) exceedance probability is thereby obtained, with the latter taken to correspond to a specified return period T_{R} via $T_{\mathrm{R}}/r_{\mathrm{i}} = 1/Q$, where $r_{\mathrm{i}} = \tau/M$. When applying this approach, the lower limit of wind speeds used in the extrapolation should be chosen judiciously so as to assure that only the tail of the distribution is suitably extrapolated.

Reference Solution A6 in Appendix A provides a spreadsheet solution for multi-hour design wind speeds on the basis of the wind data analysis procedure described above.

Example 5.1 ***Design Wind Speeds***

The wind dataset provided in Reference Solution A6 in Appendix A contains hourly wind records at a wind station over a duration of 43.22 yr. Use Reference Solution A6 to obtain the design 1-hr and 3-hr wind speeds from 140°N with a return period of 100 yr. The data has already been filtered so as to retain only directions from 130 to 150°N and wind speeds above a threshold of 35 km/hr, with consecutive rows corresponding to increasing time/date.

Solution

Since the data has already been filtered appropriately, the sets of 1-hr and 3-hr speeds shown in columns AA and AC each need to be applied to an EVA. Proceed as follows:

- Paste the 1-hr dataset in column AA (using "paste special – values") into column AK. Then paste the 3-hr dataset in column AC into column AM.
- Sort each of these columns in turn in order of decreasing speed:
  - The 1-hr speeds are found to be: 74, 72, 71, 63, … m/s.
  - The 3-hr speeds are found to be: 57, 54, 52, 50, … m/s.

- Undertake an EVA for each of these columns in turn, based on the specified duration of 43.22 yr. In each case, use 35 data points and assure that the plot is a reasonable representation of the tail of the distribution.
- In the former case, 35 data points appear appropriate; in the latter case, the lower points appear to deviate from a straight line, such that only 28 points are used.

The above procedure provides the solution as:

1-hr speed ($T_R = 100$ yr) = 85 km/hr
3-hr speed ($T_R = 100$ yr) = 61 km/hr

5.5 Wind Speed Correction Factors

In different circumstances, various correction factors may need to be applied to wind speeds. The following correction factors are described in greater detail in the Coastal Engineering Manual (2002).

5.5.1 Averaging Period

If desired, the wind speed averaged over one time interval may be converted to the *largest* wind speed averaged over a different time interval. Thus, the ratio of the *largest* wind speed averaged over a duration τ, denoted $\hat{U}(\tau)$, to a 1-hr average wind speed, denoted U_1, is taken as:

$$\frac{\hat{U}(\tau)}{U_1} = \begin{cases} 1.277 + 0.296\ \tanh[0.9 \log_{10}(45/\tau)] & \text{for} \quad \tau < 3600 \\ 1.5334 - 0.15 \log_{10}(\tau) & \text{for} \quad 3600 < \tau < 36\,000 \end{cases}$$

where τ is in seconds. This ratio may be applied when seeking, say, the 15-s gust speed (the largest wind speed averaged over 15 s) relative to a specified 1-hr speed, U_1.

Note that the above equation should not be applied when converting from, say, the *expected* wind speed averaged over a duration τ to the wind speed averaged over 1 hr, U_1. In such a case, there is no conversion needed. The reason for this is that, with reference to Figure 5.3, a randomly sampled 2-min average U_{2m} may be higher or lower than a corresponding 1-hr average, U_1. Because of this, the expected value of U_{2m} coincides with U_1.

5.5.2 Elevation

Wind speed increases with elevation, corresponding to the velocity profile in an atmospheric boundary layer adjacent to the earth's surface. The ratio of the wind speed at an elevation z (in m), denoted $U(z)$, to the wind speed at the standard 10-m elevation, denoted $U(10)$, may be taken as:

$$\frac{U(z)}{U(10)} = \left(\frac{z}{10}\right)^{1/7} \quad (z \text{ in m})$$

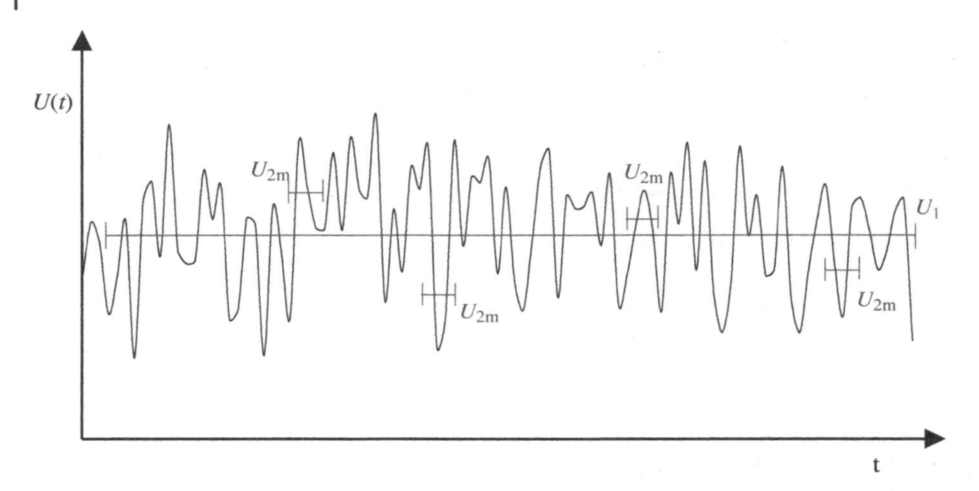

Figure 5.3 Illustration of U_{2m} occurrences in a wind record.

5.5.3 Overland to Overwater Conversion

An overwater wind speed is higher than a corresponding overland wind speed since the water surface is smooth and generates less friction than a land surface. If needed, a measured wind speed blowing over land may be converted to an equivalent wind speed blowing over water by the following approximation:

$$\frac{U(\text{overwater})}{U(\text{overland})} = 1.2$$

5.5.4 Atmospheric Stability

Atmospheric stability, which refers to atmospheric resistance to vertical air motions, is influenced notably by temperature profile with elevation. Thus, a stable atmosphere is associated with a more gradual temperature gradient or even an inversion, while an unstable atmosphere is associated with a greater temperature gradient. For winds blowing over water, atmospheric stability is influenced by the water temperature relative to the atmospheric temperature. This can affect the horizontal wind speed, such that unstable conditions may lead to greater wind speeds. A corresponding correction factor may be applied based on the following approximation:

$$\frac{U_c}{U} = \begin{cases} 0.9 & \text{for stable conditions } (\theta_a > \theta_w) \\ 1.0 & \text{for neutral conditions } (\theta_a = \theta_w) \\ 1.1 & \text{for unknown or unstable conditions } (\theta_a < \theta_w) \end{cases}$$

where U_c is the corrected wind speed, and θ_a and θ_w are the air and water temperatures, respectively.

Example 5.2 ***Wind Speed Correction Factors***
The 1-hr average overland wind speed at a certain location 10 m above the surface is 15 m/s. What are the 2-min average wind speed, the 15-s gust speed, and the 1-hr overwater wind speed at an elevation of 50 m?

Solution

Specified

U(1-hr, overland, 10 m) = 15 m/s

***U*(2-min)**

$U(\text{2-min}) = U(\text{1-hr}) = 15\ \text{m/s}$

$\hat{U}$(15-s)

$\hat{U}(\text{15-s})/U(\text{1-hr}) = 1.4$ (given by formula for $\hat{U}/U_1$)
$\hat{U}(\text{15-s}) = 21\ \text{m/s}$

***U*(overwater, 50 m)**

First, convert from overland to overwater by using:

$U\ (\text{overwater})/U\ (\text{overland}) = 1.2$

Therefore:

$U\ (10\ \text{m, overwater}) = 1.2 \times 15 = 18\ \text{m/s}$

Second, convert from 10 to 50 m by using:

$U(50\ \text{m})/U(10\ \text{m}) = (50/10)^{1/7} = 1.26$

Therefore:

$U(50\ \text{m, overwater}) = 1.26 \times 18 = 23\ \text{m/s}$

5.6 Hurricanes

Winds associated with tropical storms and hurricanes have various distinguishing features that require distinct considerations than those indicated above.

5.6.1 Tropical Cyclone Categories

A tropical cyclone is a rapidly rotating storm system characterized by a low-pressure center, strong winds, and heavy rain. A tropical cyclone may be classified in order of increasing strength as a tropical depression, a tropical storm, or a hurricane. In the Northern Hemisphere, these storms are associated with winds rotating counterclockwise. Storms with maximum wind speeds of 17 m/s (33 knots) or less are designated as tropical depressions, those with maximum wind speeds of 18–32 m/s (34–63 knots) are designated as tropical storms, and those with maximum wind speeds of 33 m/s (64 knots) or higher are designated as hurricanes.

Figure 5.4 shows a satellite view of a hurricane in the Northern Hemisphere, indicating the cloud cover that shows the hurricane eye and its counterclockwise rotation.

Figure 5.4 Satellite view of Hurricane Katrina, 2005. *Source:* NOAA/Public domain.

For the northwestern Pacific Ocean, the name "typhoon" is used in place of "hurricane," while, for the Southern Hemisphere, the terminology "tropical cyclone" is retained, regardless of strength.

5.6.2 Saffir–Simpson Scale

The Saffir–Simpson scale categorizes a hurricane into category one, two, three, four, or five, depending upon its characteristics as presented in Table 5.1. In the table, the wind speeds shown correspond to the 1-min average sustained wind speed (at an elevation of 10 m). The level of potential damage that may be caused by each category of hurricane is also indicated in the table.

5.6.3 Wind and Pressure Fields

The wind field and associated pressure field of a hurricane are sketched in Figure 5.5. The wind field is characterized by the maximum wind speed (V_{max} in the figure) and the radius of maximum wind (RMW

Table 5.1 Classification of hurricanes by the Saffir–Simpson scale.

Category	Maximum wind speed (m/s)	Minimum center pressure (mbar)	Damage
1	33–42	>980	Minimal
2	43–49	965–979	Moderate
3	50–58	945–964	Extensive
4	59–69	920–944	Extreme
5	≥70	<920	Catastrophic

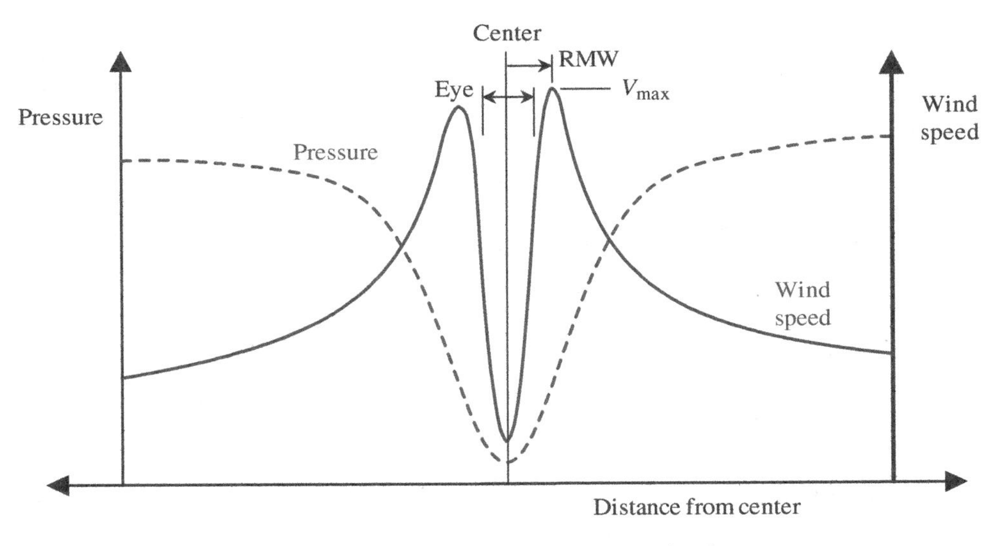

Figure 5.5 Sketch of wind speed and pressure variations across a hurricane.

in the figure), which is the radial distance from the center to the location of maximum wind speed. The slight asymmetry in the figure is associated in part with the motion of the eye itself.

Reflecting this behavior, models are available that provide the wind speed and direction in the vicinity of a hurricane. One approach to approximating the pressure field is to express the pressure p as a function of radius from the eye, r, as:

$$\frac{p - p_e}{p_a - p_e} = \exp(-\text{RMW}/r)$$

where p_a is the atmospheric pressure (far from the hurricane) and p_e is the pressure at the eye.

5.6.4 Hurricane Tracks

The track that a hurricane takes is of great importance with respect to its impact on coastal communities. In the Atlantic Ocean, hurricanes generally move slowly westward with a gradual curvature northward. Models are used to forecast a current hurricane's track and characteristics, along with indications of variability, given that an individual track is notably uncertain. NOAA's interactive portal coast.noaa.gov/hurricanes provides historical hurricane and other cyclonic storm tracks and related information. This site provides tracks of hurricanes meeting selected criteria that include tracks that have passed within a specified distance of a given location, hurricane categories, and the years in which the hurricanes occurred. As an example of the use of this site, Figure 5.6 shows the tracks of all hurricanes passing within 200 km of Miami, FL, over the period 1980–2020, with the legend indicating hurricane category (that will change along a track). Upon selecting a specific hurricane, information on maximum wind speeds and minimum pressures for a series of locations along the track may be obtained.

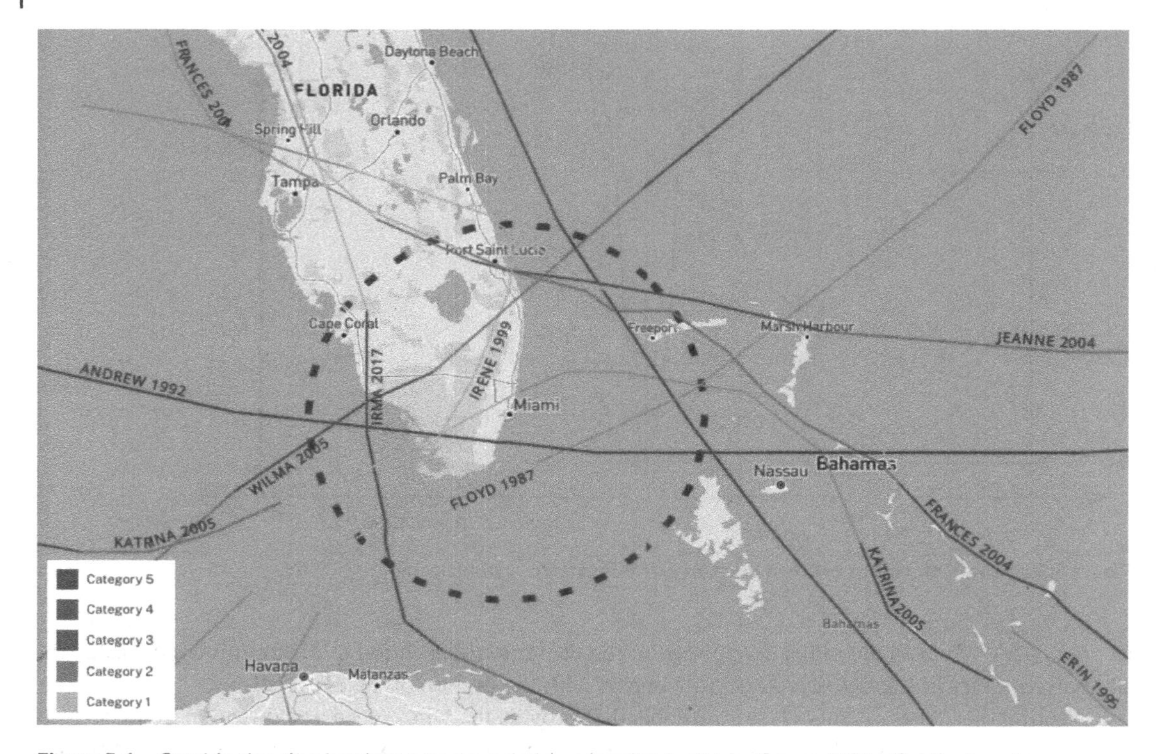

Figure 5.6 Graphic showing hurricane tracks meeting prescribed criteria. *Source:* NOAA/Public domain.

Example 5.3 ***Hurricane Wind Speeds***

Using the site coast.noaa.gov/hurricanes, obtain an image showing the tracks of all hurricanes passing within 200 km of San Juan, Puerto Rico, between 2010 and 2023. How many such hurricanes were there? Of these, which hurricane remained at category 5 continually while its eye was within 200 km of San Juan? Estimate this hurricane's most severe sustained wind speed and minimum pressure while its eye was within 200 km of San Juan.

Solution

Using the above website, select San Juan, Puerto Rico, set the search distance to 200 km, select the years 2010–2023, and deselect all but categories H1–H5.

Number of hurricanes: 6
Relevant hurricane: Irma

Select Irma and hover over its path.

Maximum sustained wind speed: 155 knots
Minimum air pressure: 915 mbar

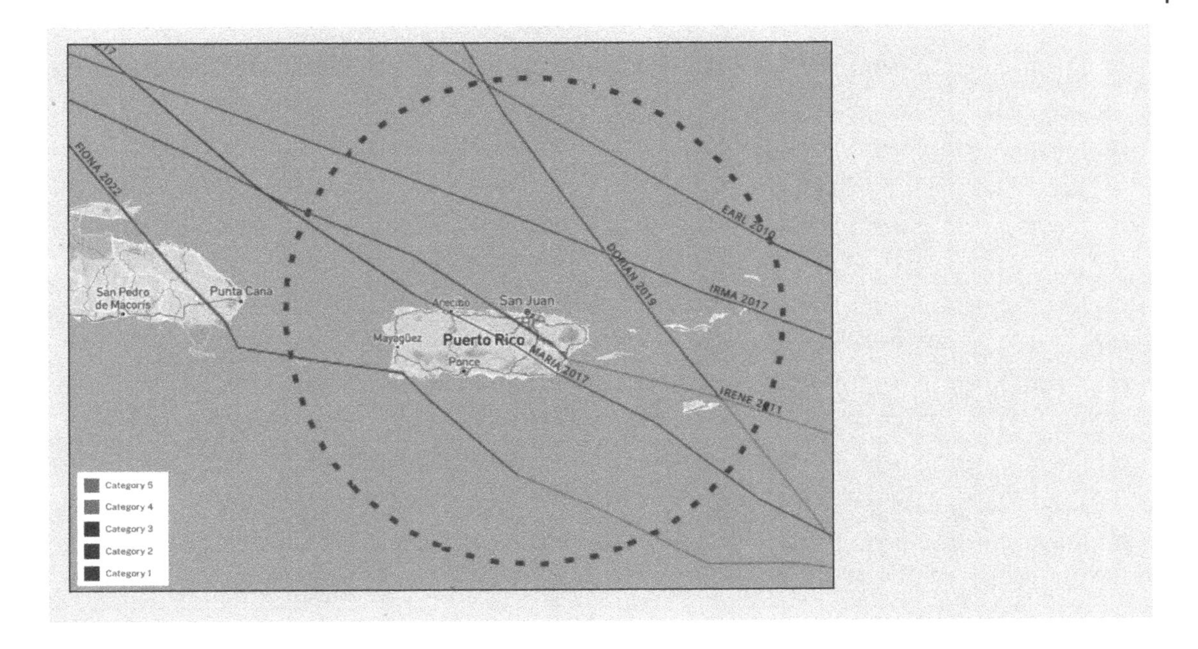

Problems

5.1 The text's companion site allows for a download of the spreadsheet *Problem Data*. The tab "Problem 5.1" contains hourly wind data for Comox, BC over the 23-yr period January 1, 2000 to December 31, 2022. Relying on Reference Solutions A5 and A6 in Appendix A, undertake an EVA to obtain the design 1-hr wind speed from 120°N with a 50-yr return period. In filtering the data, ignore wind speeds that are 39 km/hr or less and assume that data for winds from 120°N ± 10° represent the relevant wind direction. ■ Obtain also the design 3-hr wind speed following the same procedure.

5.2 The 1-hr average wind from the northeast measured 10 m above ground at Vancouver International Airport is 40 km/hr at noon on a certain day. Taking correction factors into account as may be relevant, estimate the corresponding 2-min average speed 30 m above ground at the Tsawwassen ferry terminal nearby and the corresponding 15-s gust speed 10 m above the ground at a residence on Gabriola Island. Use Google Earth to identify these locations.

5.3 By exploring the website coast.noaa.gov/hurricanes, identify the most severe hurricane that has occurred within 100 km of Chetumal, Mexico, since 1990. How close did its eye come to Chetumal? Estimate the hurricane's maximum sustained wind speed (m/s), its minimum pressure (mbar), and the corresponding wind direction (°N) at Chetumal, all at the time that it was closest to Chetumal.

6

Wave Predictions

6.1 Introduction

A determination of design wave conditions is central to many coastal engineering projects. Since these invariably refer to wind-generated waves, the ability to predict the magnitude of wind-generated waves by examining the associated wind field is of fundamental importance. The usual approach that is adopted, termed wave hindcasting, is to utilize past wind data to estimate corresponding wave conditions. Thus, the focus of this chapter is on wave hindcasting. A general outline of approaches that may be adopted is given below. This is followed by a description of a simplified approach to hindcasting in Section 6.2 and an indication of numerical hindcast and forecast models in Section 6.3.

Apart from wind-generated waves, knowledge of other forms of wave generation may be relevant with respect to certain aspects of coastal engineering. In this context, Sections 6.4 and 6.5 outline, respectively, the prediction of ship waves and laboratory-generated waves.

6.1.1 General Approaches

There are three key steps needed to obtain design wave conditions for application to a project. The first one is to undertake an extreme value analysis (EVA) so as to obtain extreme wind or wave conditions, usually corresponding to a specified return period. The second is to undertake a hindcast analysis so as to obtain deep-water waves for known wind conditions. The third is an analysis of the transformation of waves to the site. These three steps, along with optional subsidiary steps, may be undertaken in different ways and different sequences as summarized below.

Wind records. Available wind data from one or more wind stations are obtained. Usually, this is in the form of 1-hr average wind speeds and associated wind directions, taken every hour over several years, and/or may be in the form of resulting wind speed statistics for various wind speed and direction ranges.

Extreme value analysis. An EVA is applied to the wind data in order to determine design wind speeds in a form that is suitable for application to a wave hindcast. An alternative approach, which is more suited to the application of a numerical hindcast model, is to undertake a wave hindcast for a series of storms using hourly wind data within each storm and then to apply an EVA to the hindcast wave conditions for each storm. Furthermore, the EVA may be applied to deep-water wave conditions prior to determining

An Introduction to Coastal Engineering, First Edition. Michael Isaacson.

Companion website: www.wiley.com/go/coastalengineering

wave transformations to the site or, alternatively, to wave conditions at that site that take account of wave transformations for each of the storms.

Hindcast analysis. A hindcast analysis is applied so as to use wind data to estimate corresponding wave conditions. This may be either applied to wind data associated with individual storms prior to undertaking an EVA or applied to design wind conditions. It may also be undertaken by a simplified analysis utilizing available formulae and procedures or a computer model that may combine the hindcast analysis with wave transformations to a site.

Wave transformations. Wave transformations to a site, such as by shoaling, refraction, and diffraction, may be developed from deep-water design wave conditions for individual storms or for a design storm. These may be assessed by spreadsheet calculations, a standalone numerical model, or a numerical model that combines the hindcast analysis with wave transformations.

Wave records. Measured wave records at a site are invariably not available or are insufficient to be able to obtain design wave conditions directly. However, if available, wave records at one or more locations may be used to validate or calibrate the use of a wave hindcast model.

Additional parameters. Additional parameters that may be required in combination with H_s and T_p include the mean wave direction, a measure of the sea state duration, the prevailing current, and maximum water levels. Typically, H_s and T_p are used with the most severe combinations of these other parameters, but in some cases, a joint probability analysis may be warranted. As well, the wave field itself may be described more fully, so as to include maximum wave heights, associated one-dimensional or directional wave spectra, and, for waves reaching a shoreline, breaking wave and wave runup parameters.

6.1.2 Wave Generation by Wind

Prior to describing approaches to undertaking a wave hindcast analysis, it is appropriate to indicate the physics of wave generation by wind and approaches to representing wave conditions in terms of wind characteristics. These topics have been developed since the 1950s and have included descriptions of the mechanics of wave generation (taking account of turbulence in the wind field), shear stresses exerted by the wind on the sea surface (accounting for the irregular sea surface), the energy transfer from wind to waves, the subsequent energy transfer between different wave frequencies (that is a key aspect of wave growth), and energy dissipation by wind-induced wave breaking. In particular, a distinction is made between a developing sea, which refers to a sea state in which waves continue to grow as the wind persists, and a *fully developed sea*, in which waves have reached a limit for a particular wind speed regardless of fetch and duration. In the latter case, energy transfer from the wind is in balance with energy dissipation through wind-induced wave breaking.

Related to such considerations, different parametric descriptions of wave spectra in terms of the wind field have also been developed. Of particular note, these have included the original form of the Pierson–Moskowitz spectrum, mentioned earlier, which expresses a wave spectrum in terms of wind speed for a fully developed sea, and the JONSWAP spectrum, which exhibits a relatively pronounced spectral peak and is applicable to a developing sea.

The above kinds of considerations are taken into account in the development of numerical hindcast models that are used to develop wave conditions for a known, time-varying wind field. Further information on the development of such models is provided in Section 6.3.

6.2 Wave Hindcasting – Simplified Approach

A hindcast analysis may be undertaken by a simplified method based on a set of explicit formulae and procedures or by a numerical hindcast model that develops more comprehensive descriptions of the wave field, reflecting a detailed reliance on the mechanics of wave generation by wind (see Section 6.1.2).

The former approach is referred to here as a simplified approach to wave hindcasting or as a simplified hindcast analysis and has also been referred to as traditional wave hindcasting or a desktop hindcast analysis. This topic has evolved over several decades and has included a range of methods and parametric representations. A foundational version, which is referred to as the SMB method, was based on the contributions of Sverdrup and Munk in 1947, followed by those of Bretchneider in the late 1950s. The version outlined below, which has evolved from the SMB method, is based on that provided in the Coastal Engineering Manual (2002).

Relevant formulae are based on recognizing that the significant wave height H_s and peak period T_p each depend on the wind speed U, the fetch F over which the wind blows, and the duration τ for which the wind blows. However, wave growth is limited either by the fetch, provided that the duration is sufficiently long, or by duration, provided that the fetch is sufficiently long. That is, waves are either "fetch-limited" or "duration-limited" and it is possible to rely on this feature so as to simplify the relationships needed to determine H_s and T_p.

Thus, H_s and T_p are each taken to depend on U, F, and g for fetch-limited waves or on U, τ, and g for duration-limited waves. Furthermore, both cases may be combined into a single formula by the use of an "effective fetch," F_e. For the case of fetch-limited waves, F_e is equal to the actual fetch F. For the case of duration-limited waves, F_e is the fetch that would reproduce the same waves as the duration-limited waves that depend on τ. F_e may be obtained from U, F, and τ by the following formula:

$$\frac{gF_e}{u_*^2} = \text{minimum}\begin{cases} 0.00523\left(\dfrac{g\tau}{u_*}\right)^{3/2} \\ \dfrac{gF_e}{u_*^2} \end{cases}$$

The minimum requirement indicated above reflects that the waves are either fetch-limited or duration-limited. In the above equation, u_* is termed the friction velocity and may be obtained from U through

$$\frac{u_*^2}{U^2} = 0.001\,(1.1 + 0.035U)$$

where U is in m/s. If instead U is needed in terms of u_*, then the following expression may be used:

$$\frac{U}{u_*} = 29.622 - 9.078u_* + 2.8977u_*^2 - 0.4606u_*^3 + 0.271u_*^4$$

where u_* is in m/s. Once F_e is known, then H_s and T_p may be obtained from

$$\frac{gH_s}{u_*^2} = 0.0413\left(\frac{gF_e}{u_*^2}\right)^{1/2}$$

$$\frac{gT_p}{u_*} = 0.751\left(\frac{gF_e}{u_*^2}\right)^{1/3}$$

The threshold duration τ_c below which waves become duration-limited depends on the wind speed and fetch and may be obtained by equating F_e associated with F to that associated with τ. τ_c may thereby be expressed as

$$\frac{g\tau_c}{u_*} = 33.19\left(\frac{gF}{u_*^2}\right)^{2/3}$$

As a general indication of this threshold, Table 6.1 shows the threshold durations below which waves are duration-limited for various fetches and wind speeds.

It should be noted that there is a limit to the increases of H_s and T_p with increasing fetch and duration that are given by the preceding equations. This limit corresponds to the establishment of a fully developed sea state, in which F and τ are sufficiently large that H_s and T_p are limited by wind speed alone. The equations that describe fully developed conditions are as follows:

$$\frac{gH_s}{u_*^2} = 211.5$$

$$\frac{gT_p}{u_*} = 239.8$$

These equations serve as upper limits to H_s and T_p for a given wind speed.

In applying the foregoing procedure, considerations in estimating F, τ, and U include the following:

- F is the straight-line fetch corresponding to a selected wind direction as obtained from hydrographic charts, or, more commonly, Google Earth. Judgment is needed in excluding islets from limiting the fetch.
- A higher 1-hr wind speed may result in lower waves than, say, those for a lower 3-hr wind speed. Therefore, there is usually a need to consider wind speeds for different durations in order to obtain the most severe wave conditions possible. Section 5.4 indicates an approach to determining wind speeds with

Table 6.1 Threshold duration for various fetches and wind speeds.

	Threshold duration, τ_c (hr)			
Fetch (km)	$U = 10$ m/s	$U = 15$ m/s	$U = 20$ m/s	$U = 25$ m/s
5	1.7	1.5	1.3	1.2
10	2.8	2.4	2.1	1.9
20	4.4	3.8	3.4	3.1
50	8.1	6.9	6.2	5.6

different durations by suitably analyzing a dataset containing hourly wind speeds. In lieu of analyzing hourly measurements in this way, the correction factor for averaging period, given in Section 5.5, may be used to provide a multi-hour wind speed from a given 1-hr wind speed. While the latter approach is not rigorous (consider the basis for developing the corresponding formula), it is expedient to rely on this although it has been found to be somewhat conservative.

- The wind speed corresponding to a specified direction, duration, and return period may be obtained by an analysis of wind data and an EVA in the manner described in Section 5.4.
- If relevant, U may need to be adjusted by correction factors accounting for elevation, overland-to-overwater conversion, and/or atmospheric stability as indicated in Section 5.5.
- It has been pointed out in Chapter 5 that records at a wind station nearest to a site may not suitably represent the wind speed and direction acting over the relevant fetch. Estimates of such wind speed and direction differences could be taken into account in a hindcast analysis.

The Coastal Engineering Manual (2002) elaborates on and extends the above summary with respect to, for example, a consideration of shallow depths, narrow fetches, and improved wind speed correction factors relating to atmospheric stability and elevation changes.

Reference Solution A7 in Appendix A provides a spreadsheet solution reflecting the simplified hindcast analysis described above.

Example 6.1 ***Simplified Hindcast Analysis 1***

A wind with speed $U = 52$ km/hr acting over a fetch $F = 30$ km lasts for a duration $\tau = 3$ hr. Based on a simplified hindcast analysis, what are the significant wave height, the peak period, and the threshold duration below which the waves are duration-limited? Correction factors should not be applied to the wind speed.

Solution

Specified

$g = 9.80665\ \text{m/s}^2$
Wind speed $U = 52$ km/hr
$= 14.4$ m/s
Fetch, $F = 30$ km
Duration, $\tau = 3$ hr

Wave height and period

From the relevant formulae, obtain in turn

$u_* = 0.58$ m/s

gF_e/u_*^2 (fetch-limited) $= 878\,243$

gF_e/u_*^2 (duration-limited) $= 409\,400$

Therefore, duration-limited. Hence:

$$\frac{gF_e}{u_*^2}\text{(actual)} = \text{min. of above} = 409\,400$$

$$gH_s/u_*^2 = 0.0413\left(gF_e/u_*^2\right)^{1/2} = 26.4$$

$$gT_p/u_* = 0.751\left(gF_e/u_*^2\right)^{1/3} = 55.8$$

$$H_s = 26.4\,u_*^2/g = 0.9\text{ m}$$

$$T_p = 55.8\,u_*/g = 3.3\text{ s}$$

Threshold duration

From formula for τ_c:

$$\frac{g\tau_c}{u_*} = 304\,381$$

$$\tau = 17\,964\text{ s} = 5.0\text{ hr}$$

Example 6.2 ***Simplified Hindcast Analysis 2***

What steady wind speeds blowing for a duration $\tau = 2$ hr over fetches $F_1 = 8$ km and $F_2 = 12$ km in turn will generate waves with $H_s = 1.2$ m? Wind speed correction factors should not be applied.

Solution

Specified

$g = 9.80665\text{ m/s}^2$
Height, $H_s = 1.2$ m
Fetch, $F_1 = 8$ km
Fetch, $F_2 = 12$ km
Duration, $\tau = 2$ hr

Approach

The approach adopted is to assume fetch-limited waves, obtain u* for the specified F and H_s, and then use this u_* to test whether the waves are fetch-limited or duration-limited. If the waves are fetch-limited, then u_* is maintained. If the waves are duration-limited, then u_* is instead obtained from τ and H_s. Once u_* is known, this is used to obtain U.

Case 1: $F_1 = 8$ km

Assuming fetch-limited waves, the formula for H_s in terms of $F_e = F_1$ can be recast as

$$u_* = \left(\frac{1}{0.0413}\right)\sqrt{\frac{gH_s^2}{F_1}} = 1.02\text{ m/s}$$

Verify the fetch-limited assumption for this u_*:
For fetch-limited: $gF_e/u_*^2 = \frac{gF_1}{u_*^2} = 75\,808$
For duration-limited: $gF_e/u_*^2 = 0.0523\left(g\tau/u_*^2\right)^{3/2} = 95\,634$

Therefore, the fetch-limited assumption is correct.
The formula for U/u_* then gives

$$\frac{U}{u_*} = 23.19$$

$$U = 23.6\ \text{m/s}$$

Case 2: $F_2 = 12$ km

Assuming fetch-limited waves, the formula for H_s in terms of $F_e = F_2$ can be recast as

$$u_* = \left(\frac{1}{0.0413}\right)\sqrt{\frac{gH_s^2}{F_2}} = 0.83\ \text{m/s}$$

Verify fetch-limited assumption for this u_*:
For fetch-limited: $gF_e/u_*^2 = \frac{gF_2}{u_*^2} = 170\,569$
For duration-limited: $gF_e/u_*^2 = 0.0523\left(g\tau/u_*^2\right)^{3/2} = 129\,622$

Therefore, the fetch-limited assumption is incorrect. Therefore, u_* is instead obtained for the specified τ and H_s. The formulae for H_s in terms of F_e and for F_e in terms of τ can be combined to yield

$$u_* = \left[\frac{1}{0.0413 \times \sqrt{0.00523}}\ \frac{g^{1/4} H_s}{\tau^{3/4}}\right]^{4/5} = 0.93\ \text{m/s}$$

The formula for U/u_* then gives

$$\frac{U}{u_*} = 23.53$$

$$U = 21.8\ \text{m/s}$$

6.3 Wave Hindcasting and Forecasting – Numerical Models

While the discussion so far has focused on wave hindcasting, a distinction is now made between wave hindcasting and wave forecasting. A wave hindcast refers to an estimation of wave conditions based on past wind data, whereas a wave forecast refers to an estimation of wave conditions based on information on future winds, usually obtained from meteorological forecasts. Both hindcasts and forecasts are based on the same methodology and differ only with respect to the timeline of the wind information that is utilized.

6.3.1 Spectral Wave Models

Section 6.1.2 summarizes both the physics of wave generation by wind and approaches to representing wave conditions in terms of wind characteristics. Information is now provided on the development of computer models to determine wave conditions that are based on this physics. Wave hindcasts and forecasts are undertaken by the application of spectral wave models, also called wind wave models. These are based on the application of an energy balance equation that enables a description of a directional wave spectrum. The equation involves three source terms that correspond to the input of energy from the wind, the nonlinear transfer of energy within a wave spectrum, and the dissipation of energy due to white-capping. In shallower water, energy dissipation from seabed friction and from wave breaking is also accounted for.

The earliest models, referred to as first-generation models, did not account for nonlinear wave interactions, whereas second-generation models developed in the 1980s relied on a parametric representation of these nonlinear interactions. Subsequently, third-generation wave models, which are now widely used, represent all three source terms explicitly, without requiring any assumptions regarding spectral shape. The third-generation models are able to utilize a wind field with temporal and spatial variations over large areas to develop a changing directional wave spectrum and so, in turn, representative wave heights, periods, and directions. These models are highly sophisticated and entail advanced numerical and computational schemes in order to assure stability and accuracy.

Spectral wave models in deep water can be extended so as to apply to intermediate and shallow water depths, develop hindcasts with regional or global coverage, or develop wave forecasts. These are indicated in turn.

6.3.2 Extension to Intermediate and Shallow Depths

When spectral wave models are applied to intermediate and shallow depths, the energy balance equation within a model needs to be extended to take account of dissipation both by wave breaking as waves approach a shoreline and by bottom friction, possibly accounting for a muddy bottom and vegetation. They take account of wave transformations, including shoaling, refraction, diffraction, and reflection, and possibly wave-current interactions and time-varying depths due to tidal fluctuations. The application of such models typically entails the specification of bathymetric data, shoreline configurations, mesh generation, the preparation and use of wind time series inputs, and the specification of model parameters. Figure 3.25 provides an example of the computational grid and bathymetry used in such a wave hindcast/transformation model and of significant wave height contours and wave directions obtained from such a model.

6.3.3 Regional and Global Models

The spectral wave models may also be applied over very large areas so as to provide regional and even global wave climates, relying on suitable linkages to regional and global meteorological models. At these scales, models may incorporate available buoy, platform, ship, and satellite altimeter wind and wave measurements.

These models may be run over an extended period of several decades in order to develop a description of the long-term wave climate over large areas. As one example, Figure 6.1 shows hindcast results in the form of the 99th percentile annual significant wave heights around the globe. This particular hindcast was generated using a model referred to as GROW. It was based on a simulation of about 40 years of wind data and was validated against buoy, platform, and satellite altimeter wave measurements. Note that even

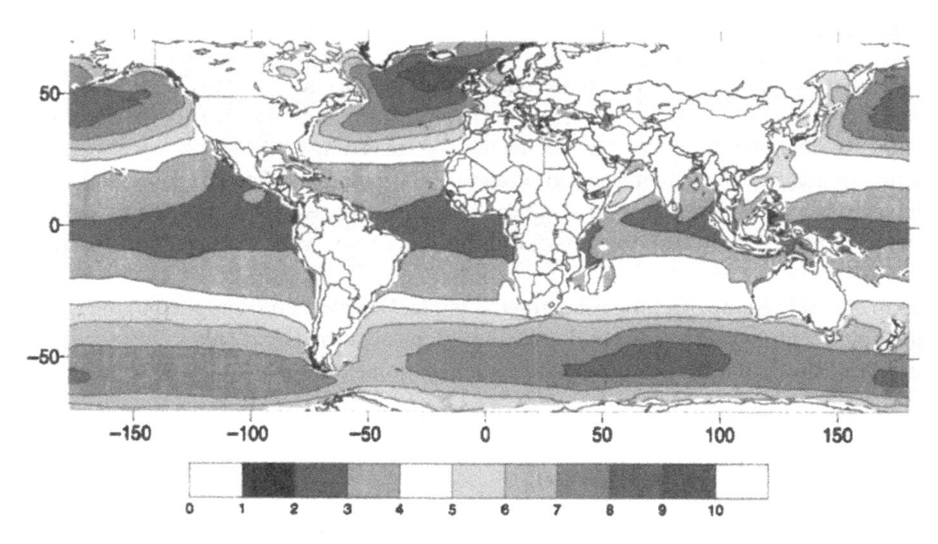

Figure 6.1 Distribution of the 99th percentile annual significant wave height obtained from a global wave hindcast. *Source:* Cox and Swail (2001). Reproduced with permission of Wiley.

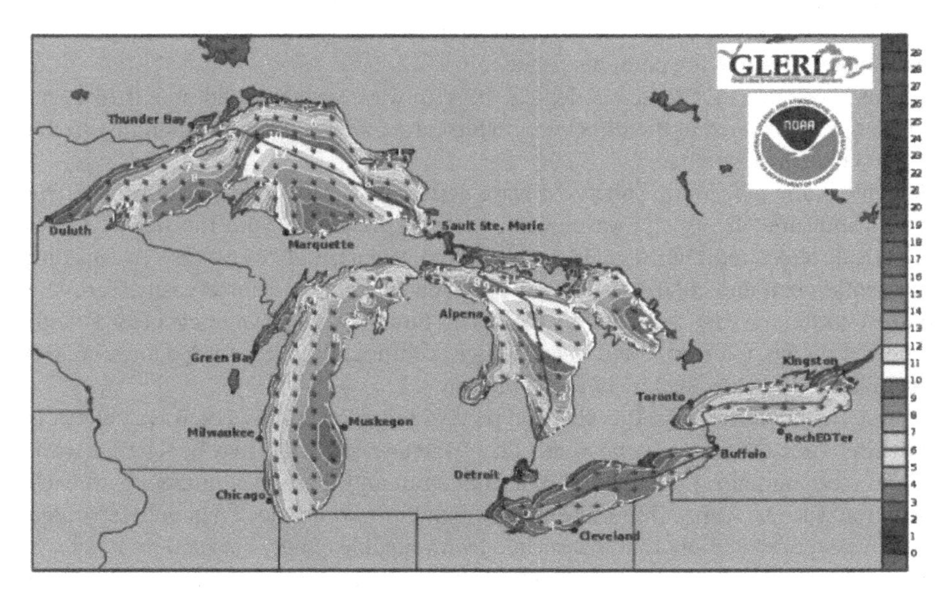

Figure 6.2 Distribution of significant wave height (in ft) and wave direction for the Great Lakes obtained from a wave forecast model. *Source:* NOAA/Public domain.

the 99th percentile annual significant wave height does not provide information on extreme waves such as those arising from cyclones and extreme storms. Instead, information on extreme waves would need to be captured by a different kind of application of such models.

6.3.4 Operational Forecasting

Spectral wave models may be combined with meteorological forecasts in order to provide operational forecasts of wave conditions extending up to 7–10 days into the future, analogous to the period over which weather forecasts are provided. Furthermore, such wave forecasts may be developed on a regional or a global basis and the models may be run at frequent intervals to provide, in effect, ongoing real-time wave forecasts. Such wave forecasts are used by the shipping industry with respect to optimizing voyage routes and by the offshore industry with respect to undertaking short-term wave-sensitive construction operations.

Figure 6.2 provides an example frame of a forecast using the model WaveWatch III applied to the Great Lakes. The figure shows significant wave heights and wave directions over the Great Lakes for a specified future time. The source website provides the corresponding animation extending up to 150 hr from the present time.

6.4 Ship Waves

In some cases, waves generated by ships and other marine vessels may require consideration, so it is of interest to consider fundamental approaches to determining such waves. The classic pattern of ship waves due to a moving vessel in deep water, sometimes referred to as a *Kelvin wave pattern*, is illustrated in Figure 6.3. Figure 6.3a provides a photograph of such a wave system, while Figure 6.3b shows a simulation of such a system when caused by a moving point disturbance.

In order to explore this wave pattern, Figure 6.4 depicts the ship-wave system based on a theoretical formulation. In the figure, x is distance in the direction of ship travel, with $x = 0$ corresponding to the ship location, and y is distance orthogonal to x.

As noted from Figures 6.3 and 6.4, a ship-generated wave system comprises two sets of waves. The solid lines in Figure 6.4 indicate "diverging" waves whose crests spread out from the line of vessel motion – these are relatively short and high. The dashed lines indicate a set of "transverse" waves that cross the line of vessel motion perpendicularly – these are relatively long and are not as easily seen. The wave system is contained within a wedge indicated by the dotted lines that make an angle of 19.5° with the line of vessel motion as shown, referred to as a *Kelvin wedge*. Additionally, at the wedge location, the wave crests make an angle of 54.7° with the line of vessel motion.

In practice, a wave system may differ from this classical pattern in various ways. The wave pattern is modified at reduced depths when the waves are no longer deep-water waves. There may be separate wave systems for the bow and stern and also for each hull of a catamaran and these may interact with each other, differently at different speeds. Also, a ship wake usually includes a pronounced region of separated flow, in part generated by propeller motion, that is associated with notable energy dissipation.

Transverse waves travel at the vessel speed V, while diverging waves travel more slowly (in a direction perpendicular to their crests) because of their different direction. As a general indication of these, the wave speed c is given as

$$c = \begin{cases} V & \text{for transverse waves} \\ 0.82\,\mathrm{V} & \text{for diverging waves} \end{cases}$$

In either case, the wave length L and wave period T may be obtained from the linear dispersion relation.

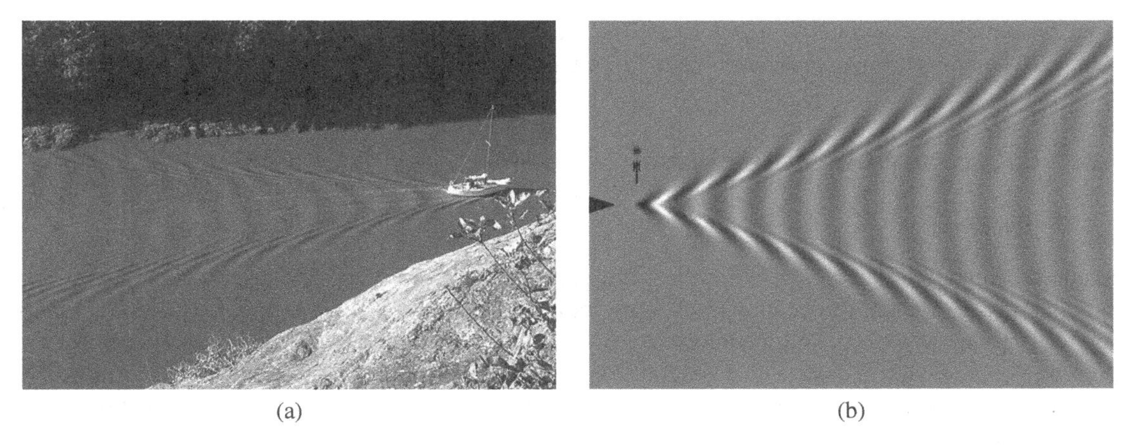

Figure 6.3 Views of a ship-generated wave system. (a) Photograph of ship-generated waves, (b) simulation of a wave system due to a moving point disturbance. *Sources:* (a) Arpingstone/Wikimedia Commons/Public domain. (b) L3erdnik/Wikimedia Commons/CC BY SA 4.0.

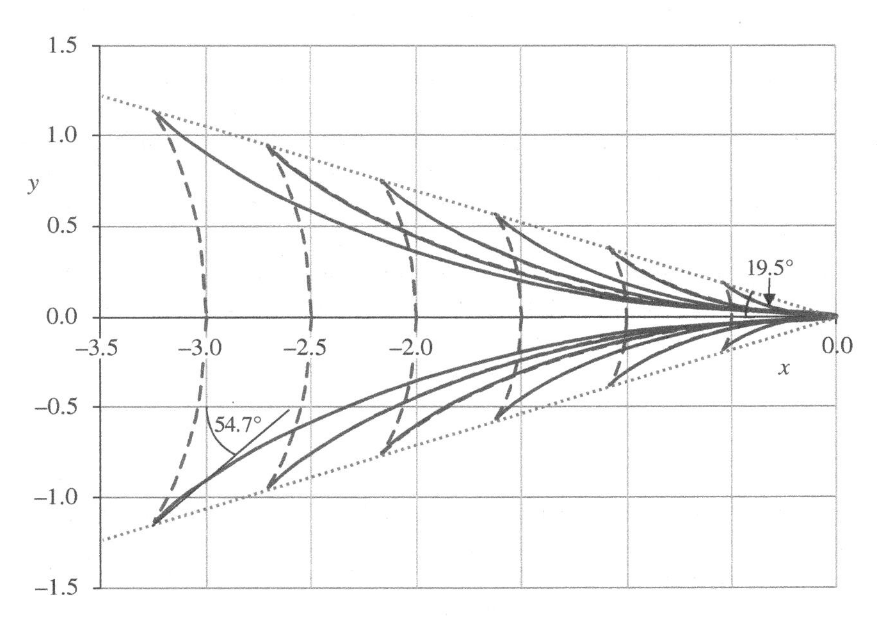

Figure 6.4 Definition sketch of a ship-generated wave system.

The wave heights of the wave system are sensitive to hull characteristics and the vessel speed, so simple parametric formulae for the wave height over a range of conditions appear not to be available. Typically, the wave height increases rapidly with speed over certain speed ranges, due to interference effects between components of the generated wave system. Wave energy density diminishes for portions of the wave system further behind the vessel, where the wave system becomes wider. It may thereby be shown that wave height is inversely proportional to $\sqrt{r}$, where r is distance from the line of vessel motion.

6.5 Laboratory-Generated Waves

6.5.1 Overview

In coastal engineering, laboratory modeling often involves a wave flume, with a wave generator at one end that is used to generate a unidirectional wave train, or a wave basin with a segmented generator that is used to generate an oblique or directional wave field. Such facilities are described in Section 12.5. For the present, the prediction of a wave field generated by prescribed motions of a wave generator is outlined. A variant of the approaches described here may be used to simulate landslide-generated waves under some circumstances.

For a wave flume, the wave generator may be driven sinusoidally to generate periodic waves, or it may be computer-controlled so as to generate random waves or transient waves. For a wave basin with a segmented generator, the programming of the generator motions needs to take account of phase differences between adjacent segments so as to develop oblique waves with a prescribed direction or, by extension, a specified directional wave spectrum.

For the more fundamental case of sinusoidal motions of a generator in a wave flume intended to create a progressive wave train, it is noted that the generator's horizontal velocity variation over depth would need to match continually the time-varying hyperbolic cosine profile of the progressive wave train in order to satisfy the boundary condition at the generator surface. Since this is not possible, the progressive wave train that is generated is supplemented by a series of *evanescent waves* adjacent to the generator, each corresponding to a standing wave whose amplitude decays with distance from the generator. Evanescent waves are of more general interest, since they occur in other situations involving wave flows around barriers. In such cases, while they do not affect the flow in the far field, they do need to be taken into account with respect to wave runup and wave pressures adjacent to a barrier.

6.5.2 Wavemaker Theory

In the following, wavemaker theory is used to predict progressive and evanescent waves arising from the oscillatory motions of a generator in a wave flume. Dean and Dalrymple (1991) provide an outline of wavemaker theory for a range of situations. The linear theory describing waves resulting from the oscillatory motion of a wave generator in a flume is now summarized.

The generator typically corresponds to a piston or a paddle hinged at or near the floor. The solution is developed here using complex notation in the manner indicated in Section 2.6.4 so as to illustrate an application of such a representation.

Consider the two-dimensional wave flow in a flume resulting from oscillatory motions of a wave generator located at $x = 0$. In the following, the case of an oscillating piston is treated, with its time-varying displacement given by $\xi(t) = X \sin(\omega t)$, where X is the displacement amplitude. Using complex notation, this may be written as $\xi(t) = iX \exp(-i\omega t)$.

Small amplitude oscillations are assumed so that the boundary value problem may be linearized. The velocity potential of the resulting flow satisfies the Laplace equation within the fluid region, the two linearized free surface boundary conditions along $z = 0$, the boundary condition on the flume bottom $z = -d$, a radiation condition indicating that the waves propagate in the x direction far from the generator, and a generator surface boundary condition at $x = 0$.

A general solution for ϕ and η that satisfies the various boundary conditions is given as follows:

$$\phi = \left[A_o \cosh(k_o s) \exp(ik_o x) + i \sum_{n=1}^{\infty} A_n \cos(k_n s) \exp(-k_n x) \right] \exp(-i\omega t)$$

$$\eta = \frac{i\omega}{g} \left[A_o \exp(ik_o x) + i \sum_{n=1}^{\infty} A_n \exp(-k_n x) \right] \exp(-i\omega t)$$

where $s = z - d$ and the coefficients A_n need to be determined from the generator-surface boundary condition.

The above solution corresponds to the superposition of a progressive wave train (subscript o) with wave number k_o, consistent with the term $\exp[i(k_o x - \omega t)]$, and a series of evanescent waves ($n = 1, 2, 3, \ldots$), which are standing waves whose amplitudes decay with distance from the generator, consistent with the terms $\exp(-k_n x)$. Thus, k_1, k_2, k_3, ... correspond to the wave numbers of successive evanescent wave modes. If needed, the corresponding wave lengths are given as $L_o = 2\pi/k_o$ and $L_n = 2\pi/k_n$ for $n = 1, 2, 3, \ldots$

k_o and k_n may be obtained from a set of dispersion relations given as

$$k_o d \tanh(k_o d) = \omega^2 d/g$$

$$k_n d \tan(k_n d) = -\omega^2 d/g \quad \text{for } n = 1, 2, 3, \ldots$$

The first of these is the usual dispersion relation for a progressive wave train, while the second of these has multiple solutions corresponding to increasing values of k_n. Of these, the longest wave corresponding to k_1 becomes negligible at a distance of $2d$–$3d$ from the wave generator. That is, the progressive waves alone persist beyond the vicinity of the generator.

In determining $k_n d$, an iterative solution based on Newton's method may be obtained using

$$X_{m+1} = X_m - \frac{X_m \tan(X_m) + A}{\tan(X_m) + X_m \sec^2(X_m)}$$

with $A = \omega^2 d/g$ and $X_o = n\pi$. $k_n d$ is then taken as X_m when m is sufficiently large.

A solution for the coefficients A_n may be obtained by an application of the generator surface boundary condition. Consistent with linear theory, this condition is linearized so as to equate the horizontal fluid velocity to the horizontal generator velocity along $x = 0$. That is,

$$\frac{\partial \phi}{\partial x} = \omega X \exp(-i\omega t) \qquad \text{at } n = 0$$

The coefficients A_n are thereby obtained as

$$A_o = -\frac{i\omega X}{k_o} \left[\frac{4 \sinh(k_o d)}{2k_o d + \sinh(2k_o d)} \right]$$

$$A_n = -\frac{i\omega X}{k_n} \left[\frac{4 \sin(k_n d)}{2k_n d + \sin(2k_n d)} \right] \qquad \text{for } n = 1, 2, 3, \ldots$$

The above solution may be used to develop a transfer function T_F that relates the progressive wave amplitude to the generator displacement amplitude:

$$T_F = \frac{4\tanh(k_o d)\sinh(k_o d)}{2k_o d + \sinh(2k_o d)}$$

This methodology can be extended to obtain the transfer function for other generator configurations such as a paddle hinged at the flume bottom. Also, transfer function formulae for different frequencies may be appropriately superposed to obtain a random generator motion that is needed to give rise to a prescribed wave spectrum. Likewise, it is possible to program the motions of a segmented generator that comprises a series of short, independently controlled generators, so as to obtain a prescribed three-dimensional regular or random wave field.

Problems

6.1 A wind with speed $U = 60$ km/hr blows over a fetch $F = 30$ km in a given direction adjacent to a coastal site. However, the corresponding duration is not known. Estimate the maximum possible H_s and T_p values for this site. What is the minimum duration of the 60 km/hr wind that is needed for such waves to occur? Correction factors should not be applied to the wind speed.

6.2 A steady wind with speed $U = 40$ km/hr blows for a duration $\tau = 1$ hr across a lake that has a fetch $F = 6$ km in the wind direction. If x is distance along the fetch measured from the upwind end of the lake, develop a plot showing the variation of the significant wave height H_s with x. What is H_s in the middle of the lake, $x = 3$ km? Above what value of x are the waves duration-limited?

6.3 ■ A ship is moving in deep water at a speed of 4 m/s parallel to a straight shoreline that is 80 m away. How long after the ship passes point A on the shore do the ship waves generated at A reach the shore? What are the wave speed and the wave direction relative to the shoreline of the diverging waves along the edge of the Kelvin wedge when they reach the shore? The maximum wave height is found to be 0.9 m at the shore. How far from the shore should the ship travel (at the same speed) for the maximum wave height to be 0.6 m instead?

6.4 ■ Laboratory waves generated by a piston generator have a period $T = 2.0$ s. The water depth $d = 0.8$ m. What are the wavelengths of the progressive waves and the first two modes of the evanescent wave system, L_0, L_1, and L_2, respectively? What is the ratio of the first-mode evanescent wave amplitude at a distance d from the generator face to the progressive wave amplitude?

7

Long Waves, Water Levels, and Currents

To this juncture, the text has largely considered wind-generated waves. Attention is now given to other kinds of waves for which the length is large relative to the water depth, $L \gg d$. These include tides, tsunamis, landslide-generated waves, flood waves in rivers, storm surge, harbor oscillations, and other low-frequency water level fluctuations. Initially, long wave theories that may be used to describe such waves are outlined. This is followed by specific considerations relating to different kinds of long waves. The outline relating to long waves is followed by descriptions of the assessment of water levels with regard to potential coastal flooding and the consequences of coastal flooding. Finally, a brief outline of currents as may be relevant to coastal engineering projects is provided.

7.1 Long Wave Theories

Prior to describing the various forms of long waves and related water level fluctuations, theoretical approaches to describing long waves, which do not rely on the limitations of regular wave theories (i.e. periodic waves on water of constant depth), are initially summarized. Linearized and nonlinear long wave theories that incorporate a wide range of situations are outlined in turn below.

7.1.1 Linearized Long Wave Theory

The shallow (long) wave limit of linear wave theory was provided in Chapter 2, whereas extensions that are applicable to a wider range of conditions are now outlined. The key assumptions made are that the wave height is small, $H \ll L$, d, so that nonlinear terms may be omitted (as with linear wave theory) and that the depth is small relative to the wave length, $d \ll L$. The latter assumption implies that the horizontal velocity u is uniform over depth and that the pressure p is hydrostatic with respect to the instantaneous water surface elevation η, i.e. $p = \rho g(\eta - z)$. However, other assumptions for a regular wave train may be relaxed, so that the unsteady or transient propagation of long waves may be simulated, and as well the water depth need not be constant. This allows for the modeling of transient long waves over an uneven bathymetry, as in the case of tidal propagation in estuaries.

As a starting point, consider a unidirectional, unsteady flow in the x direction for the case of constant depth d. $u(x, t)$ and $\eta(x, t)$ are the two variables to be determined. Under these assumptions, the

An Introduction to Coastal Engineering, First Edition. Michael Isaacson.

Companion website: www.wiley.com/go/coastalengineering

governing equations for u and η may be obtained from the continuity and momentum equations, given, respectively, as

$$d\frac{\partial u}{\partial x} = \frac{\partial \eta}{\partial t}$$

$$\frac{\partial u}{\partial t} + g\frac{\partial \eta}{\partial x} = 0$$

These can be combined to yield the one-dimensional wave equations for constant depth:

$$\frac{\partial^2 \eta}{\partial t^2} - c^2\frac{\partial^2 \eta}{\partial x^2} = 0$$

$$\frac{\partial^2 u}{\partial t^2} - c^2\frac{\partial^2 u}{\partial x^2} = 0$$

where $c = \sqrt{gd}$. The general solution for η can be written in the form:

$$\eta = f_1(x - ct) + f_2(x + ct)$$

A numerical solution for particular situations may be obtained by finite difference methods or the method of characteristics.

The appeal of the linearized long wave theory is that it may readily be extended. In particular, it may be extended to two dimensions in the horizontal plane so as to solve for $u(x, y, t)$, $v(x, y, t)$, and $\eta(x, y, t)$ and to variable depths $d(x, y)$ by using $q = ud$ in place of u. It may also be extended to take account of other effects such as time-varying depths, bottom friction, wind stresses, and Coriolis effects. Numerical models based on the foregoing may be used to examine long wave propagation under various circumstances.

7.1.2 Nonlinear Long Wave Theories

Nonlinear long wave theories based on the Boussinesq-type equations that incorporate variable depths, nonperiodic time variations, and other forms of wave transformation were outlined in Section 3.11.3. Since the modern foundation of such methods was laid in the 1960s and 1970s, corresponding numerical models have been developed to a high level of sophistication, such that they are now able to simulate a wide range of effects including wave shoaling, refraction, diffraction, variable depths, relative depths that approach deep water conditions, transient boundaries, seabed friction, partially absorbing or reflecting boundaries, wave nonlinearities, and wave breaking.

Likewise, the numerical implementation of such models has included the development of various finite difference and finite element formulations, the use of nested grids representing the horizontal domain, and different time-stepping formulations, with close attention given to assuring numerical stability and accuracy. Additionally, different kinds of boundary conditions may be incorporated, including incident wave conditions along a portion of the model boundary, a specification of a moving boundary or seabed (for instance with respect to tsunami generation), and boundary conditions within the domain or along its edges associated with barriers and with partial absorption or reflection. An application of such a model to tsunami propagation is illustrated in Section 7.3.

7.2 Tides

7.2.1 Introduction and Historical Development

Tides refer to the oscillatory variation of sea level due to gravitational interactions between the sun, moon, and earth. As a more extreme example, the tidal range, which is the elevation difference between low tide and high tide, extends up to about 17 m within the Bay of Fundy, NS. The elevation fluctuation with time depends on the latitude, the time of the year, and the specific location. The oscillatory variation may take on different forms, as illustrated in Figure 7.1.

That is, a diurnal tide exhibits one minimum and one maximum a day, a semidiurnal tide exhibits two minima and two maxima a day, and a mixed tide exhibits two highs of unequal height (higher high water and lower high water) a day and two lows of unequal height (lower low water and higher low water) a day.

The historical development of tidal predictions is summarized. The equilibrium theory of tides, described by Newton in 1687, considers a hydrostatic response of the water body to gravitational effects of the sun, moon, and earth, assuming a spherical earth with a uniform water depth. Despite these

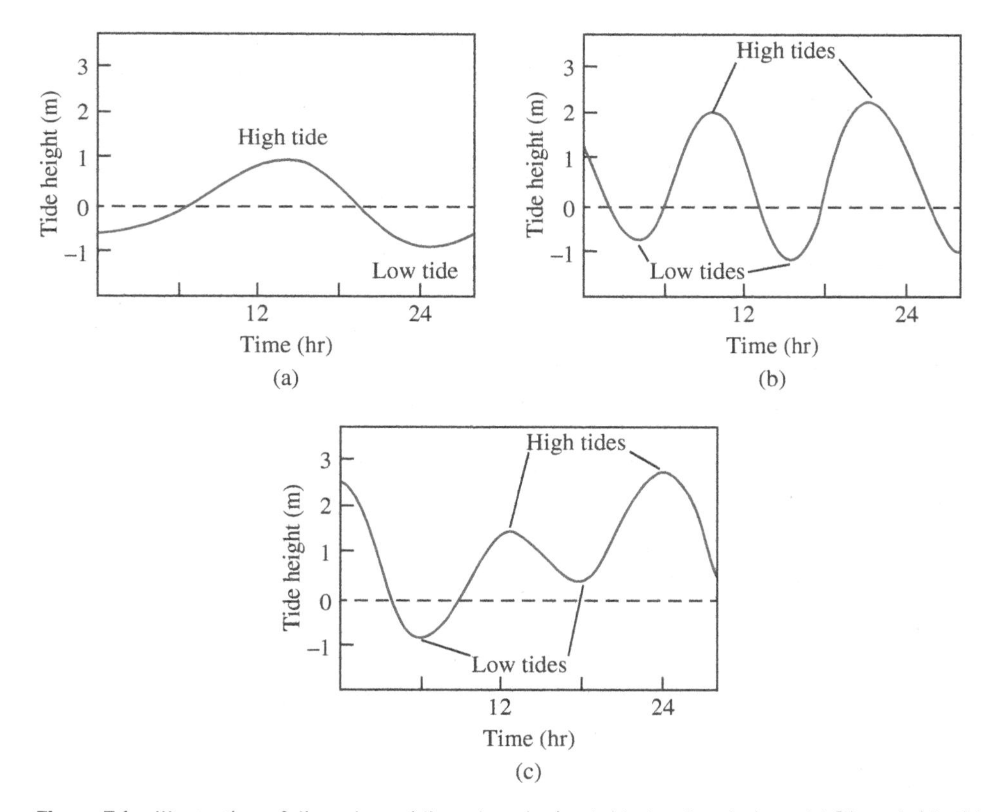

Figure 7.1 Illustration of diurnal, semidiurnal, and mixed tide level variations. (a) Diurnal tide, (b) semidiurnal tide, (c) mixed tide. *Source*: NOAA/Public domain.

restrictive assumptions, this simple theory does predict some gross features of tides, including two low tides and two high tides each day. Subsequently, the dynamic theory of tides, developed by Laplace in 1775, takes account of a rotating earth, the effects of land masses, and water depth variations. As a result, this was able to predict tidal fluctuations more closely. Subsequent advances, including the harmonic representation of tidal records, were not made until the 1860s and beyond.

7.2.2 Glossary

A glossary relating to kinds of tides and tidal flows includes the following:

- ***diurnal tide***: single high-water level and low-water level each tidal day
- ***semidiurnal tide***: two high-water levels and two low-water levels of approximately the same elevations each tidal day
- ***mixed tide***: contains elements of both diurnal and semidiurnal tides, with two high-water levels and two low-water levels with unequal elevations each tidal day
- ***spring tide***: large tidal fluctuations when the sun and moon are aligned with the earth (corresponding to a new moon or a full moon)
- ***neap tide***: small tidal fluctuations when the sun and moon are orthogonal to each other when viewed from the earth (corresponding to the first and last quarters with respect to phases of the moon)
- ***flood tide***: rising water level
- ***ebb tide***: falling water level
- ***king tide***: a nonscientific term used to describe very high tides
- ***tidal day***: 24 hr, 50 min, corresponding to the duration of the earth's rotation with respect to the moon passing over the same meridian
- ***tidal range***: vertical distance between high tide level (called high water) and low tide level (called low water)
- ***tidal datum epoch***: duration used in the determination of certain reference tide levels, recognizing that tidal fluctuations include a component that repeats every 18.6 yr

A list of acronyms relating to sea levels and tide levels includes the following:

MHHW: mean higher high water – average of all higher high-water levels from 19 consecutive years of predictions

MLLW: mean lower low water – average of all lower low-water levels from 19 consecutive years of predictions

MSL: mean sea level – mean elevation of the sea surface over an extended region (e.g. an entire sea) and over an extended duration (e.g. a year)

MWL: mean water level – average observed water level at a given location over a given duration (e.g. during a storm)

HAT: highest astronomical tide – highest predicted tide at a given location. (Tidal fluctuations include a component that repeats every 18.6 yr, so that the highest astronomical tides recur with that period.)

The above list of tide levels and their definitions correspond to US usage. In Canada, MHHW is referred to as HHWMT, higher high water mean tide, and MLLW is referred to as LLWMT, lower low water mean tide. In addition, the following are used:

HHWLT: higher high water large tide – average of highest high-water levels, one from each of 19 consecutive years of predictions

LLWLT: lower low water large tide – average of lowest low-water levels, one from each of 19 consecutive years of predictions

In other countries, other high and low tide levels may instead be used. For example, in Australia, New Zealand, and the UK, high tide levels that are used are referred to as mean high-water springs (MHWS), mean high-water neaps (MHWN), and mean higher high water (MHHW). Likewise, low tide levels that are used are referred to as mean low-water springs (MLWS), mean low-water neaps (MLWN), and mean lower low water (MLLW).

Finally, specific alternatives to some definitions, such as those relating to the tidal datum epoch, have been adopted by different organizations.

7.2.3 Prediction of Tide Levels

Tidal predictions are based on assuming a harmonic series for the tidal elevation η relative to the mean sea level:

$$\eta = \sum_i A_i \cos(\omega_i t - \delta_i)$$

where A_i, ω_i, and δ_i are, respectively, the amplitude, angular frequency, and phase angle of each constituent. The amplitudes and phases depend on location, such that the combination of various constituents gives a unique signature of a tidal record for that location. Each harmonic is associated with a different mechanism, some more pronounced than others. The eight most prominent constituents are indicated in Table 7.1.

Table 7.1 Principal constituents of tidal records.

	Symbol	Period (hr)	Relative amplitude
Semidiurnal	M2	12.42	100.0
	S2	12.00	46.6
	N2	12.66	19.1
	K2	11.97	12.7
Diurnal	K1	23.93	58.4
	O1	25.82	41.5
	P1	24.07	19.3
Long period	Mf	327.86	17.2

The strongest component (M2 term) is termed the primary lunar semidiurnal constituent. There are other lower frequency harmonics that are not shown, including one with a period of 18.6 yr, referred to as a lunar nodal cycle, that contribute to the HAT at a particular location.

7.2.4 Vertical Datums

Tidal elevations need to be expressed relative to a specified vertical datum. The vertical datum that is adopted is either a tidal datum, based on a reference tidal elevation at a particular location, or a datum used in surveying and global positioning applications.

The tidal datum used in the US is MLLW and so differs from one location to the next. In Canada, the tidal datum is referred to as Chart Datum (CD) in hydrographic charts and tide tables, usually corresponding to LLWLT, and so differs from one location or chart to the next.

Datums that are used in surveying and global positioning are referred to as orthometric or gravimetric datums that rely on the geoid, which is a gravitational equipotential representation of the earth's surface, and so correspond to an idealized mean sea level over the globe. Other datums are referred to as three-dimensional or ellipsoidal datums that rely on a reference ellipsoid, which is an ellipsoidal representation of the earth. Of these, the latter datums involve six geocentric parameters with respect to origin and orientation, so that they serve simultaneously as vertical and horizontal datums.

Overall, the vertical datums used most commonly in the US are NAVD 88 (which is an orthometric datum) and NAD 83 (which is a three-dimensional datum), while the vertical datums used most commonly in Canada are CGVD2013 (also referred to as CGG2013) and a previous vertical datum CGVD28, with either one referred to as Geodetic Datum (GD).

Datum conversion. Conversion between different datums for particular locations in the US can be accomplished using NOAA's Vertical Datum Transformation website via the link below.[1] For locations in Canada, the conversion between CD and GD may be obtained most effectively by referring to values for LLWLT, MSL, and/or HHWLT relative to GD, available from the Fisheries and Oceans Canada (DFO) CAN-EWLAT data portal using the link below,[2] and relative to CD available from hydrographic charts or from DFO's Canadian Tide and Current Tables website using the link below.[3]

7.2.5 Tidal and Bathymetric Data

In coastal engineering projects, there is usually a need to access tidal and bathymetric data. Bathymetry is the study of water depths, so that bathymetric data describes depth variations over a designated area, which can, in effect, generate a submarine relief map.

Such data include reference tide levels (e.g. MHHW), predicted (i.e. astronomical) tides, measured tides (differences from predicted tides are associated primarily with storm surge), hydrographic charts, and bathymetric data. A range of such information is available, although somewhat different formats are used.

For the US, tidal data may be accessed from NOAA's Tides and Currents website using the link below.[4] For a selected location, this site includes links that provide reference tide levels (referred to as "datums"), tidal predictions that include annual tide tables, harmonic constituents, measured extreme tide levels, and projected extreme values for various return periods.

---

1 vdatum.noaa.gov/vdatumweb/.
2 dfo-mpo.gc.ca/science/oceanography-oceanographie/accasp-psaccma/can-ewlat-ocanee/index-eng.html.
3 charts.gc.ca/publications/tables-eng.html.
4 tidesandcurrents.noaa.gov.

For Canada, tidal data may be accessed from DFO's Tides, Currents, and Water Levels website using the link below.[5] For a selected location, this site includes links that provide annual tide and current tables (including information on reference tide levels and on "secondary" ports) and historical measurements of hourly tide levels.

There are usually corresponding sites in other countries that provide tide tables or other forms of tidal prediction and/or data on past tidal measurements. In addition, short-term tidal predictions are generally available worldwide from many apps, programs, and company sites.

Hydrographic charts, also referred to as nautical charts or navigational charts, usually provide reference tide levels and show bathymetry for designated areas. A hydrographic chart may be available as a paper chart, a Raster Navigational Chart, which is an electronic image of a paper chart, and an Electronic Navigational Chart, which contains additional information on different features, obtained by clicking on those features. Hydrographic charts may be obtained from a wide range of suppliers. Additional information on access to charts is available from NOAA's Office of Coastal Survey via the link below[6] for the US and from DFO's Nautical Charts and Services website via the link below[7] for Canada.

Finally, information on access to digital bathymetric data is available via the link below[8] for the US and via the link below[9] for Canada.

7.2.6 Tidal Bores

At locations with large tidal ranges, a flood tide propagating up a river may result in a tidal bore, which refers to a wave formation at the front of the tidal flow. A bore may take various forms. Most often, a tidal bore occurs as a single continual breaking wave front, analogous to a hydraulic jump, while at some locations it may occur as an undular bore that takes the form of a series of unbroken waves. Figure 7.2 illustrates both these forms. Figure 7.2a shows the world's largest tidal bore at the mouth of the Qiantang River adjacent to Hangzhou Bay in China. This bore may reach heights of up to 9 m and propagate at

Figure 7.2 Views of tidal bores. (a) Tidal bore on the Qingtang River, China, (b) undulating bore on the Petitcodiac River, Moncton, NB. *Sources*: (a) EditQ/Wikimedia Commons/CC BY SA 4.0. (b) Larry/Flickr/CC BY 2.0.

5 tides.gc.ca.
6 nauticalcharts.noaa.gov.
7 charts.gc.ca.
8 ncei.noaa.gov/products/seafloor-mapping.
9 data.chs-shc.ca/dashboard/map.

speeds up to 40 km/h. Figure 7.2b shows an undular bore, with modest wave breaking, on the Petitcodiac River in Moncton, NB.

Example 7.1 ***Predicted Tides***

A. **US Location**. Using the website tidesandcurrents.noaa.gov, what is MHHW at Friday Harbor, WA, relative to the datum MLLW and relative to the datum NAVD 88? What was high tide on December 1, 2024, relative to MLLW and at what time did it occur?

B. **Location in Canada**. Using the appropriate 2024 Tide Tables for Canada, what is HHWLT at Gibsons, BC, relative to Chart Datum (CD) taken as LLWLT? What was higher high water on December 1, 2024, relative to CD and at what time did it occur? Using the CAN-EWLAT portal, what is HHWLT at Gibsons relative to the datum CGVD2013?

Solution

A. US Location

Using the above website, enter WA, Friday Harbor, More Data, Datums, set Datum to MLLW; read off:
MHHW: 2.364 m MLLW
Set datum to STND; read off:
MHHW: 2.538 m NAVD 88
Enter Tides/Water Levels, NOAA Tide Predictions; set date range, units, datum, data only:
high tide: 2.59 m MLLW
at time: 6:56 am

B. Location in Canada

Gibsons is referred to as a secondary port. The 2024 Tide Tables indicate that the reference port is Point Atkinson, and that, relative to Point Atkinson:
HHWLT: +0.1 m
at time: −1 min
For Point Atkinson:
HHWLT: 5.0 m CD
December 1 higher high: 4.7 m CD
at time: 6:49 am
Therefore, for Gibsons:
HHWLT: 5.0 + 0.1 = 5.1 m CD
December 1 higher high: 4.7 + 0.1 = 4.8 m CD
at time: 6:49–0:01 = 6:48 am
From the CAN-EWLAT portal for Gibsons:
HHWLT: 2.15 m CGVD2013

7.3 Tsunamis

7.3.1 Introduction and Examples

A tsunami is a rare event corresponding to a series of high-speed, long-period water waves caused by a rapid change to the water boundary, with the potential of severe inundation at exposed shorelines.

A tsunami may be generated by a distant or local earthquake, a submarine landslide, or an above-water landslide entering the water – or even by volcanic eruptions or large meteors. Tsunamis generated in the open ocean are the most well known and are now considered. Landslide-generated waves generated by either a submarine landslide or an above-water landslide entering the water are described later in this section.

Because of their long periods, tsunamis are shallow-water waves, even in the deep ocean, so that the wave speed is given by $c = \sqrt{gd}$, where d is the water depth. The wave height and surface slope are very small in deep water so that they are then imperceptible. As waves approach the shore, the wave height increases significantly. These waves may remain unbroken and appear as a rapidly rising tide, or they may break and appear as a bore advancing toward the shore. A tsunami comprises a series of individual waves. The initial indication of an arriving tsunami may be a receding water level corresponding to a leading wave trough, followed by flooding as a wave crest propagates inland. Engineering activities relating to tsunamis include the development and use of warning systems and emergency response procedures, predictions of travel times and wave runup elevations, the definition of potential inundation zones, and possibly shoreline protection measures.

While there are many examples of catastrophic tsunamis worldwide, comments are now provided on two specific tsunamis.

1964 Alaska Tsunami, March 27, 1964. This tsunami was generated by a 9.2 magnitude earthquake whose epicenter was about 130 km east of Anchorage, AK, and it resulted in notable damage and over 100 deaths as far south as California. In Canada, the tsunami resulted in over 400 homes being damaged or destroyed at Port Alberni, BC. Port Alberni is at the head of the 60 km long Alberni Inlet, such that resonance within the inlet led to wave amplification and an extended duration of resonant wave activity. Figure 7.3 shows the tidal record at the time of the tsunami's arrival, including both the measured elevation and the predicted astronomical tidal variation. The highest waves, with a runup of about

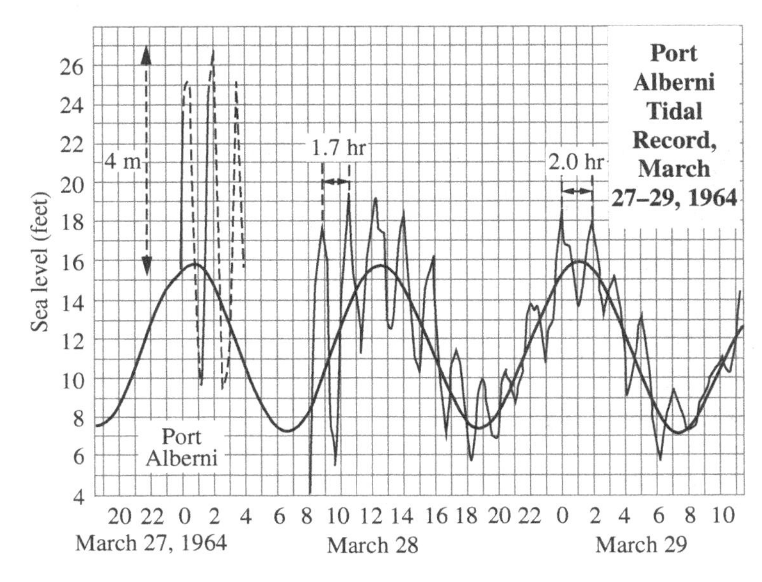

Figure 7.3 Tide record at Port Alberni, BC, during the arrival of the 1964 Alaska tsunami. *Source*: Fine et al. 2009. Reproduced with permission of Springer.

3 m, were coincident with high tide, leading to an especially high degree of inundation. The figure also indicates that the tsunami had a period of 1.7–2.0 hr and demonstrates the persistence of the tsunami oscillations.

2004 Indian Ocean Tsunami, December 26, 2004. This tsunami, sometimes referred to as the Boxing Day tsunami, was generated by a 9.3 magnitude earthquake whose epicenter was about 80 km off the coast of North Sumatra, and it led to some 230 000 deaths in 11 countries. Figure 7.4 highlights various aspects of the tsunami, including the initial drawdown of water associated with a wave trough initially arriving at a shore (Figure 7.4a), the approach of the first tsunami crest in the form of a bore or broken wave (Figure 7.4a), subsequent flooding as the tsunami propagated inland (Figure 7.4b), an indication of

Figure 7.4 Aspects of the 2004 Indian Ocean tsunami. (a) Initial drawdown and approaching bore, (b) subsequent flooding, (c) example of aftermath, (d) tsunami travel times. *Sources*: (a) NOAA/Public domain. (b) Kristin Shorten/Wikimedia Commons/CC0 1.0. (c) AusAID/Wikimedia Commons/CC BY 2.0. (d) NOAA/Public domain.

the aftermath (Figure 7.4c), and its predicted travel times (Figure 7.4d). The latter indicates, for example, only two hours for the tsunami's arrival in Sri Lanka.

7.3.2 Tsunami Modeling

Tsunami modeling is based on the methods described in Section 7.1.2 with respect to shallow-water waves, notably the Boussinesq-type models that incorporate past or potential ground movements associated with designated events, as well as a discretization of the propagation area. As an example of the use of such a model, Figure 7.5 illustrates some aspects of a model used to examine a tsunami resulting from a severe earthquake arising in the Cascadia subduction zone off the west coasts of the US and Canada. The images in the figure were generated using the Boussinesq wave model FUNWAVE-TVD

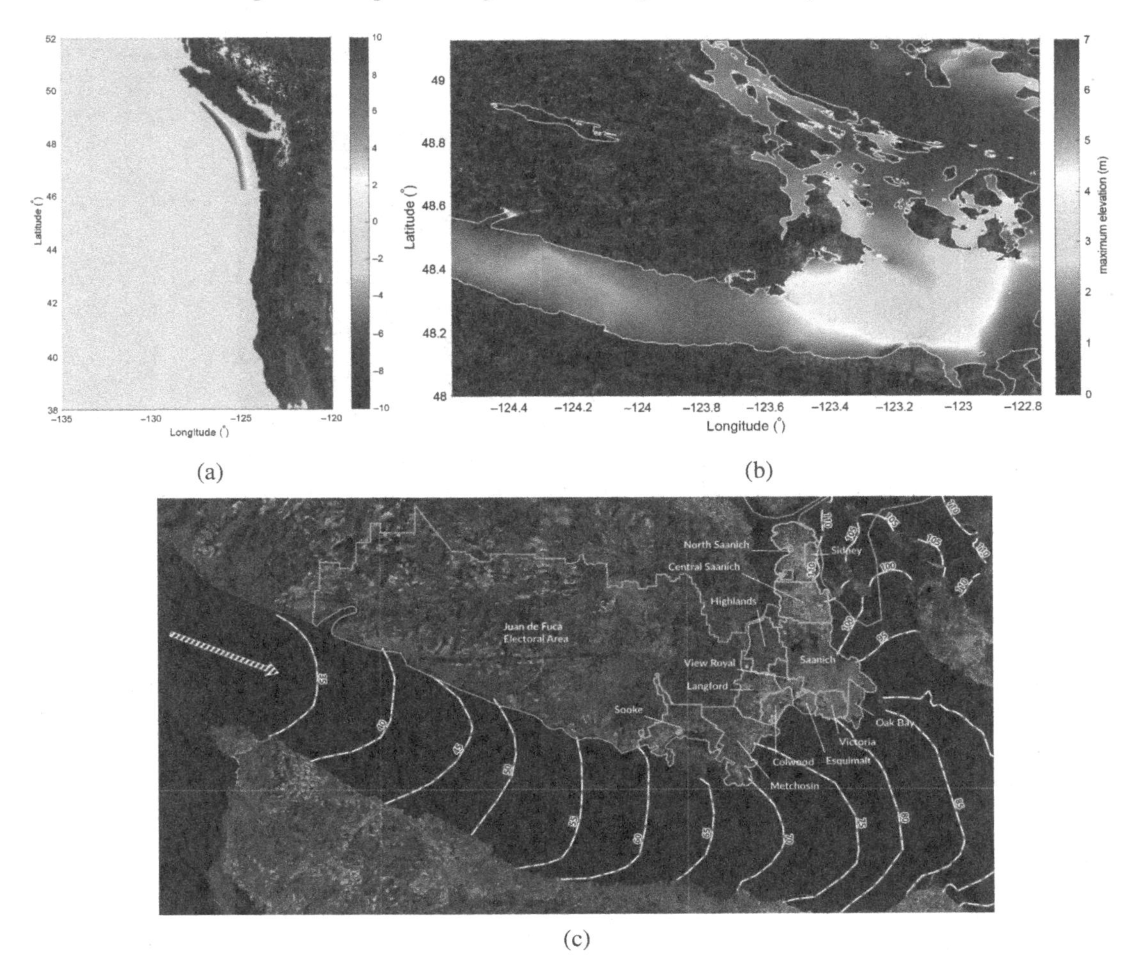

Figure 7.5 Example application of a tsunami model. (a) Vertical ground displacement, (b) maximum water surface elevation, (c) travel times. *Source*: Reproduced with permission of the Capital Regional District (BC).

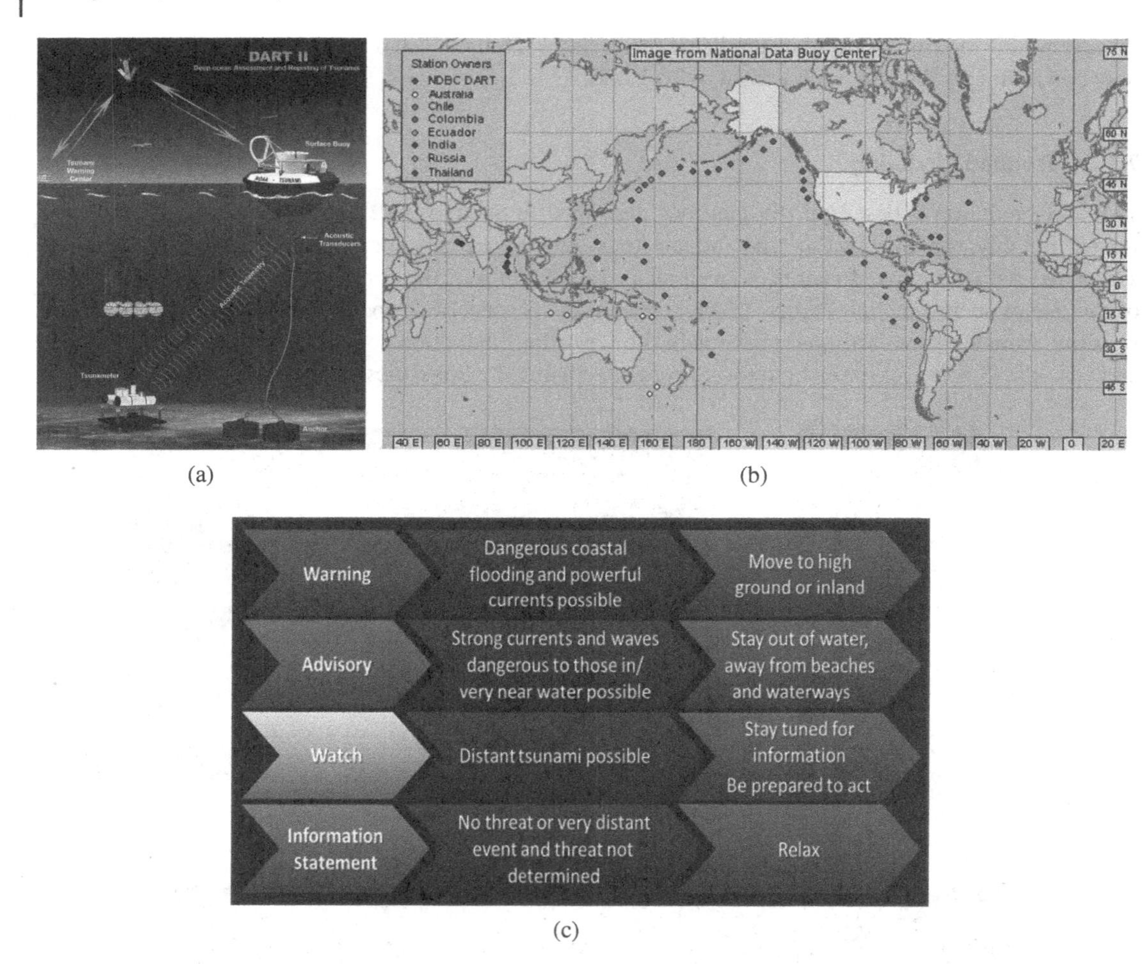

Figure 7.6 Illustration of aspects of a tsunami warning system. (a) Sensors and communications, (b) sensor locations, (c) types of tsunami alerts for the US. *Source*: NOAA/Public domain.

developed at the University of Delaware. The figure illustrates in turn the vertical ground displacement of the tsunamigenic event relating to the northern portion of the Cascadia subduction zone (Figure 7.5a), the resulting maximum water surface elevation in the Juan de Fuca Strait (Figure 7.6b), and the corresponding tsunami travel times (Figure 7.5c).

7.3.3 Tsunami Runup Predictions

Since a tsunami is a rare event (e.g. once in a few hundred years to once in over a thousand years), there is insufficient information to develop runup values corresponding to a specified AEP or return period. Instead, a different approach is adopted, relying on a consideration of different past or potential tsunami sources/events, which typically have estimated return periods of the order of 500 yr or more, along with an assessment of the resulting runup values at a site. Such an assessment would normally be based on a modeling study or make reference to previous studies and reports. As may be relevant, the sources/events may include tsunamis originating in the open ocean, locally generated tsunamis arising from an earthquake at shallow crustal depths, submarine landslides associated with unstable sediments on a submarine slope, and a sudden slide associated with a steep, unstable coastline. The runup from a locally-generated shallow-depth event is usually significant only very close to the source.

7.3.4 Tsunami Warning Systems and Emergency Management

A tsunami warning system is used to detect tsunamis and issue warnings to prevent loss of life and damage to property. Such a system comprises a network of sensors, including seismic sensors to detect seismic activity, tide gauges, buoys, and seafloor pressure sensors used to detect distant wave activity that verify the presence of a tsunami and a communications infrastructure to issue alarms to evacuate coastal areas. Figure 7.6a provides a schematic of a sensor network, Figure 7.6b indicates sensor locations across the Pacific Ocean and beyond, and Figure 7.6c summarizes tsunami alert levels as adopted in the US.

7.3.5 Landslide-Generated Waves

Similar to ocean tsunamis, landslide-generated waves may arise from a submarine landslide or an above-water landslide entering the water. Submarine landslides are possible when seabed material along a submarine slope is unstable, so that such a slide may be triggered by an earthquake. Examples in BC include two small tsunamis caused by submarine landslides at the mouth of the Kitimat River in 1974 and 1975, and as well the potential of a larger tsunami due to a major slide on the western front of the Fraser River delta triggered by an earthquake. Small tsunamis may also be generated in the immediate vicinity of a large, sudden landslide associated with a steep, unstable coastline. Localized wave activity arising from debris avalanches is also possible.

The magnitude of landslide-generated waves depends primarily on the volume of water displaced by the slide and by the slide duration and is usually significant only very close to the slide. Landslide-generated waves may be modeled in the same manner as described for tsunamis. Predictions of landslide-generated waves may also be made on the basis of simplified assumptions of slide geometry and an analogy with a wave generator in a flume for the case of unidirectional wave propagation or a wave generator within the wall of a wave basin for the case of wave directional spreading (see Section 6.5).

7.4 Long Wave Oscillations

Seiche (usually pronounced sāysh or seysh) are long-period, low-amplitude standing wave oscillations in enclosed or partially enclosed bodies of water such as lakes, harbors, and inlets. They are initiated by forcing from wind setup or wave setup, tsunamis, or earthquakes, with the oscillations continuing after the forcing mechanism has ceased. A fundamental illustration of seiche corresponds to the standing waves in basins considered in Section 3.4. Numerical methods based on wave diffraction theory may be used to predict seiche for basins of arbitrary shape and constant or variable depth.

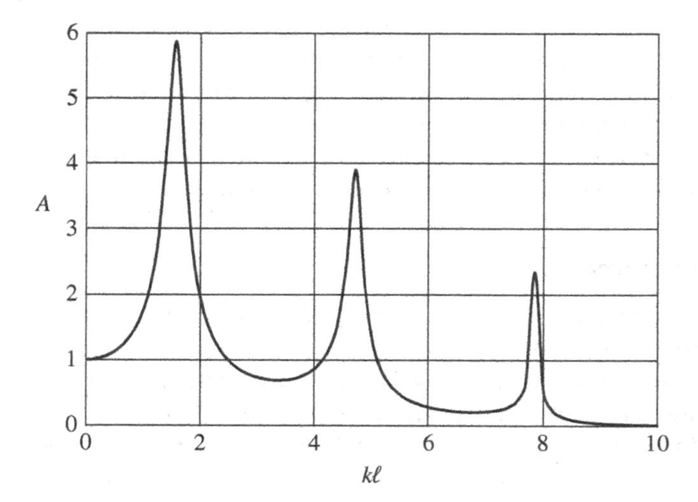

Figure 7.7 Amplification factor as a function of relative frequency for a harbor.

As one example, Lake Erie, which has an average depth of 19 m and a length of 388 km, exhibits seiche with a fundamental period of 14 hr 18 min, whereas the simple formulae given in Section 3.4.2 provide this period for an effective length of 351 km – which is only slightly smaller than the actual length.

In the case of a harbor, long-period oscillations may coincide with the natural periods of moored ship motions, leading to excessive motions and mooring system failure. Thus, mooring analyses for moored ships may require a consideration of potential long-period oscillations in harbors. The analysis of harbor resonances or related laboratory tests is usually carried out at different wave frequencies so as to obtain an amplification factor A, which is the wave amplitude at an inner point of the harbor relative to the incident wave amplitude, as a function of relative frequency $k\ell$, where k is the wave number and ℓ is the harbor length. This is illustrated in Figure 7.7, which shows resonance peaks associated with the various modes.

7.5 Storm Surge

Storm surge refers to an increase in MWL above the astronomical tide level because of a storm and so excludes wave-by-wave fluctuations. Storm surge is a major consideration with respect to elevated water levels due to hurricanes causing flooding, notably along the south and east coasts of the US. As illustrated in Figure 7.8, some waterfront houses along parts of the US east coast are elevated so as to minimize damage from potential storm surge.

However, storm surge is not restricted to hurricane-prone areas and is almost always significant at most coastal locations.

Storm surge may be obtained as a deep water or regional component, with subcomponents due to wind setup, pressure setup, and longer-term fluctuations that may or may not be associated with a particular

Figure 7.8 Elevated waterfront houses at Outer Banks, NC. *Source*: Pixabay/Public domain.

storm system and a local component due to wind setup associated with nearby winds acting over shallower depths in the vicinity of a site. These are considered in turn below. Wave setup is considered in Section 7.6.

7.5.1 Regional and Local Storm Surge

Because weather systems have a length scale of hundreds of kilometers, regional storm surge affects a broad region without being localized to a particular site. This component includes longer-term fluctuations in MWL that are not associated with a particular storm (see Section 7.5.4).

Regional storm surge may be estimated on the basis of previous studies involving analyses of long-term tidal records at a series of tide stations. This approach also captures longer-term fluctuations in MWL that have been mentioned above. Alternatively, storm surge – including local storm surge – may be estimated by an analysis of tidal measurements at a particular tide station. Predicted tidal records (astronomical tides only) are subtracted from measured tidal records (astronomical tides plus storm surge), and then an extreme value analysis is applied to obtain the residual water level extremes for selected return periods.

Local storm surge includes wind setup and wave setup. In some cases, the latter is included within wave runup, so that local storm surge would then be taken to correspond to local wind setup alone.

Apart from a reliance on tidal records, regional and local storm surge may also be obtained from numerical modeling in conjunction with a specified bathymetry, shoreline profile, and time-varying wind and pressure fields.

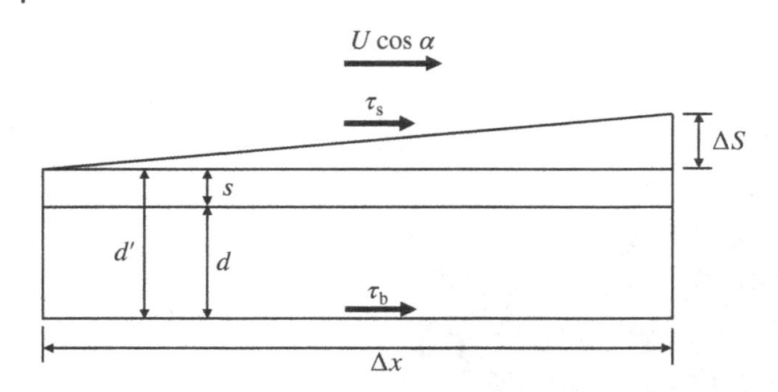

Figure 7.9 Definition sketch of wind setup.

7.5.2 Wind Setup

Wind setup is a rise in local water level approaching a shoreline due to a sloping water surface caused by wind stresses acting on the water surface. Estimates of wind setup may be obtained by referencing relevant studies, by numerical modeling, or from the use of simple formulae based on a steady wind acting over an assumed fetch with a uniform or variable depth. Approaches based on simplified numerical modeling and simple formulae are summarized below.

A governing equation for the variation of setup s with distance x toward the shore may be developed with reference to the definition sketch for a segment of the water column of length Δx, as shown in Figure 7.9.

In the figure, s is the setup at a location x, d is the still water depth without setup, $d' = d + s$ is the local water depth including setup, U is the wind speed, and α is the angle between the direction in which the wind is blowing and the x direction. A corresponding force balance provides

$$(\tau_s + \tau_b)\Delta x = \frac{1}{2}\rho g(d' + \Delta s)^2 - \frac{1}{2}\rho g d'^2$$

where τ_s is the shear stress at the water surface, τ_b is the shear stress at the bottom associated with an induced current, and Δs is the incremental increase in setup over an incremental horizontal distance Δx in the direction normal to the shore.

The shear stress τ_s in the x direction may be expressed in terms of a wind drag coefficient C_d as $\tau_s = \rho_a U^2 C_d \cos\alpha$, where ρ_a is the density of air. Equivalently, by using a density-scaled drag coefficient $\kappa_s = (\rho_a/\rho)C_d$, τ_s may instead be expressed in terms of the water density as $\tau_s = \rho U^2 \kappa_s \cos\alpha$. Furthermore, some versions of such a formulation are based on the assumption that $\tau_b = 0.1\tau_s$, which may be accounted for by defining an effective drag coefficient $\kappa = 1.1\kappa_s$ so as to incorporate the effects of τ_b. Otherwise, τ_b effects may be omitted by taking $\kappa = \kappa_s$.

The above expression for τ_s may be combined with the force balance equation to develop an expression for the incremental storm surge Δs over a length Δx. This is,

$$\frac{\Delta s}{d'} = \sqrt{1 + \frac{2\,\kappa\, U^2 \cos\alpha\, \Delta x}{g d'^2}} - 1$$

This equation may be approximated to provide the setup slope ds/dx as

$$\frac{ds}{dx} = \frac{\kappa\, U^2 \cos\alpha}{gd'}$$

The empirical drag coefficient κ depends on meteorological conditions and the roughness of the water surface as influenced by wind speed, and on whether τ_b effects are accounted for or not. One approximation for κ_s that is used is

$$\kappa_s = 1.21 \times 10^{-6} + 2.25 \times 10^{-6} \times (1 - 5.6/U)^2$$

where U is in m/s. Thus, when τ_b is accounted for, the largest value of the effective drag coefficient κ is 3.8×10^{-6}, although values in the range $2.2–3.2 \times 10^{-6}$ are more typically used.

The above equations may be solved numerically to obtain s as a function of x for a given seabed profile $d(x)$, starting from a point offshore at which $s = 0$ and leading to the setup s_o at the shoreline. For a particular site and storm, the calculation may be carried out at a series of times (e.g. every hour), taking account of changes in wind speed and direction and tide (that affects depths) in order to obtain the development of wind setup with time during a storm.

As a further simplification, if a uniform depth $\overline{d}$ is used and $s \ll \overline{d}$, then the equation for setup slope may be integrated directly to estimate the total wind setup s_o at the shoreline associated with a uniform wind acting over a distance X:

$$s_o = \frac{\kappa\, U^2\, X \cos\alpha}{g\overline{d}}$$

Finally, in a narrow lake or enclosed water body with a fixed volume of water, the setup at the downwind end must be compensated by setdown at the upwind end, so that the setup at the downwind end may be approximated by the above equation but with a factor of 0.5 included.

Example 7.2 ***Wind Setup***

A lake is approximated as a long, narrow rectangular basin with an average depth $d = 6$ m and length $X = 8$ km. What is the wind setup at the downwind end due to a wind that blows at an angle of 10° with the lake axis and has a speed of 90 km/h? Assume that $\kappa = 3.2 \times 10^{-6}$.

Solution

Specified

$g = 9.80665\ \text{m/s}^2$
$U = 90\ \text{km/h}$
$= 25.0\ \text{m/s}$
$X = 8.0\ \text{km}$
$\overline{d} = 6.0\ \text{m}$
$\alpha = 10°$
$\kappa = 3.2 \times 10^{-6}$

Setup

For a lake, setdown at the upwind end implies that a factor of 0.5 needs to be included, so that the relevant formula for setup is

$$s_o = \frac{0.5\,\kappa\, U^2\, X\, \cos\alpha}{g\bar{d}}$$

Therefore:

$$s_o = 0.13\ \mathrm{m}$$

7.5.3 Pressure Setup

The pressure component of storm surge, sometimes referred to as barometric setup or barometric surge, is a rise in local water level caused by low air pressure near the center of a storm or storm system relative to the ambient or atmospheric pressure further away. A corresponding estimate of pressure setup is

$$\Delta s = \frac{p_a - p}{\rho g}$$

where p_a is the ambient or atmospheric pressure away from the storm, usually taken to correspond to a standard atmosphere, and p is the low pressure associated with the storm. Note the following conversions of units of pressure: 1 standard atmosphere (atm) = 1013.25 millibars (mbar), and 1 mbar = 1 hPa = 0.1 kPa. As an indication of the magnitude of pressure setup, a storm system with a low pressure of 964 mbar relative to a standard atmosphere gives rise to a pressure setup of about 0.5 m.

7.5.4 Long-Term Fluctuations

Longer-term fluctuations in MWL that are not associated with a particular storm may also occur and are included in regional storm surge estimates. These arise because of changes in ocean currents or temperature that are, for example, associated with El Niño and La Niña events and with Coriolis effects.

Coriolis effects cause currents to tend to veer to the right in the Northern Hemisphere, so that a current flowing parallel to shore and moving from left to right when viewed from the shore would then veer toward the shore, leading to a setup component. A simple estimate of the setup slope $\mathrm{d}s/\mathrm{d}x$ due to Coriolis effects may be obtained from

$$\frac{\mathrm{d}s}{\mathrm{d}x} = \frac{Uf}{g}$$

where U is the magnitude of the coastal current parallel to the shore and f is a Coriolis parameter defined as $f = 2\omega \sin\phi$. Here, ω is the angular velocity of the earth and ϕ is the latitude of a location. $\omega = 0.2625$ rad/hr, which leads to $f = 1.458 \times 10^{-4} \sin\phi$, with f in Hz.

7.5.5 Features of Hurricane Storm Surge

The two major components of storm surge are usually wind setup and pressure setup. The location of their maxima relative to a hurricane's center is depicted in Figure 7.10.

The magnitude of storm surge at a particular location depends on the distance from the hurricane landfall location, the local bathymetry and topography, and hurricane characteristics that include maximum wind speed, forward speed, approach angle, and minimum pressure.

In assessing the influence of approach direction, it is noted that the largest storm surge occurs as the highest speed winds blow directly onshore and that these occur (in the Northern Hemisphere) in the "right-front quadrant," i.e. to the right of the hurricane eye when viewed toward the shore. That is, the most severe storm surge arises when a hurricane approaches a shoreline normally and occurs to the right of the eye as the hurricane makes landfall. Likewise, when a hurricane moves parallel to the shore, the winds blowing directly onshore and the resulting storm surge are less severe. It is also noted that the highest wind speeds blowing offshore give rise to wind setdown, a lowering of the water level, possibly leading to the seabed becoming exposed.

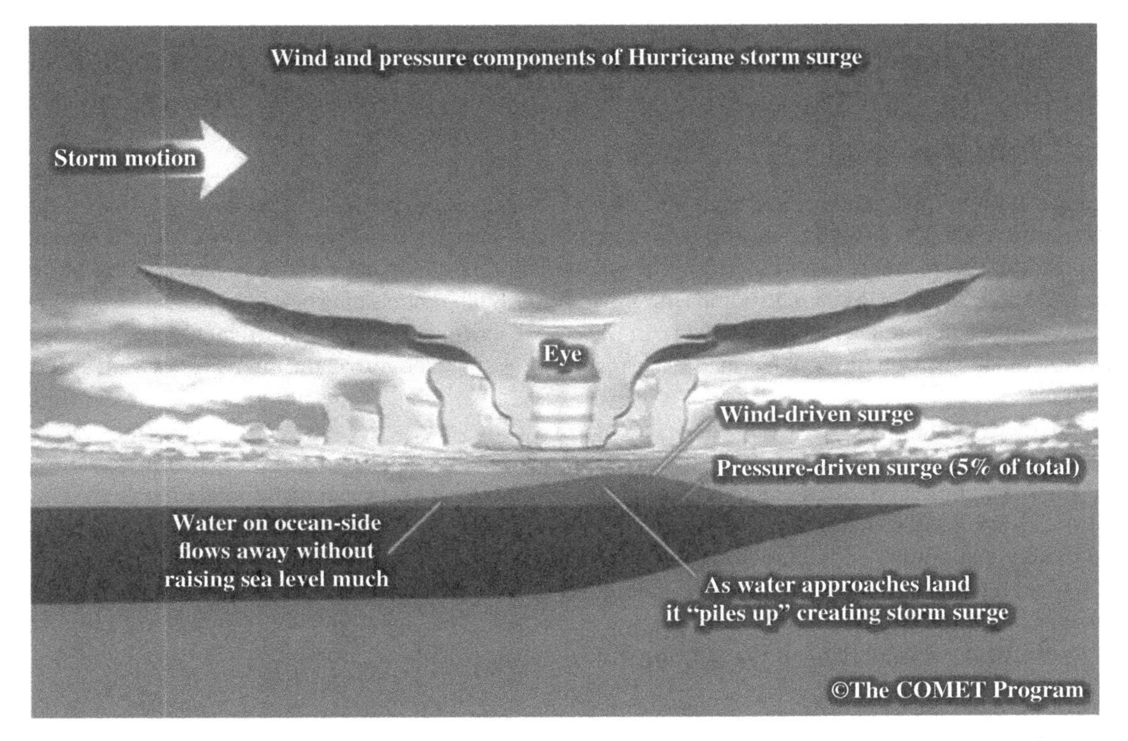

Figure 7.10 Depiction of wind setup and pressure setup components of hurricane storm surge. *Source*: ©2024UCAR. Reproduced with permission of UCAR.

Figure 7.11 Example application of a numerical model of storm surge inundation. *Source*: NOAA/Public domain.

7.5.6 Storm Surge Modeling

A range of numerical models have been developed for storm surge predictions, generally based on the methods described here with respect to wind setup and pressure setup. The application of such models includes the specification of a numerical grid of a horizontal domain, along with a specification of the wind and pressure field, bathymetry, and shoreline profile. Such models may be extended to take account of tides, wave setup, and overland flooding. In the case of storm surge due to hurricanes, the wind and pressure field may itself be modeled so as to correspond to a specified hurricane intensity, forward speed, and track.

Such models may be used to develop short-term storm surge forecasts along a coastline for an approaching hurricane, usually relying on alternate scenarios since changes in a hurricane's intensity and track are uncertain. These models may also be combined with probabilistic modeling to develop storm surge risk maps that determine storm surge vulnerability for different locations along a coastline and inland on account of overland flooding.

Figure 7.11 illustrates example results of numerical modeling stemming from the model SLOSH developed by the National Weather Service within NOAA. The figure shows storm surge elevations associated with the passage of an example hurricane and a particular shoreline.

7.6 Wave Setup

Wave setup refers to an increase in the MWL in the surf zone after waves have broken, associated with changing wave characteristics as the broken waves propagate toward the shore over decreasing depths.

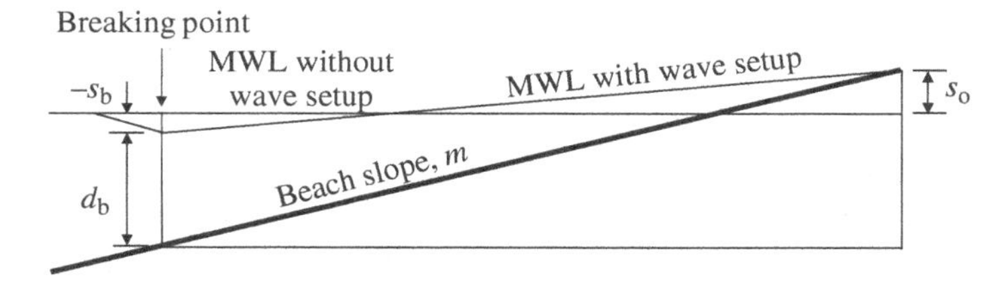

Figure 7.12 Definition sketch of wave setup.

Wave setup may be estimated distinctly or may be included within wave runup and then excluded from storm surge. Alternative formulae for wave setup are available. At the most fundamental level, the wave setup associated with a uniform beach slope is indicated in the definition sketch given in Figure 7.12.

Within the surf zone, the wave setup slope is given by

$$\frac{ds}{dx} = \frac{m}{1 + (8/3\gamma_b^2)}$$

where m is the beach slope and $\gamma_b = H_b/d_b$ is the breaker index, taken as 0.8. However, wave setdown at the breaker point is also present, given approximately by $s_b/d_b = -0.04$. Taking this into account, the wave setup at the shore is given approximately by $s_o/d_b = 0.19$.

Wave setup causes the surf zone to be wider than would otherwise be the case. Based on the geometry shown in Figure 7.13, the width of the surf zone w is given by

$$w = (d_b + s_b + s_o)/m = 1.15d_b/m$$

This is seen to be about 15% wider than would be the case without wave setup.

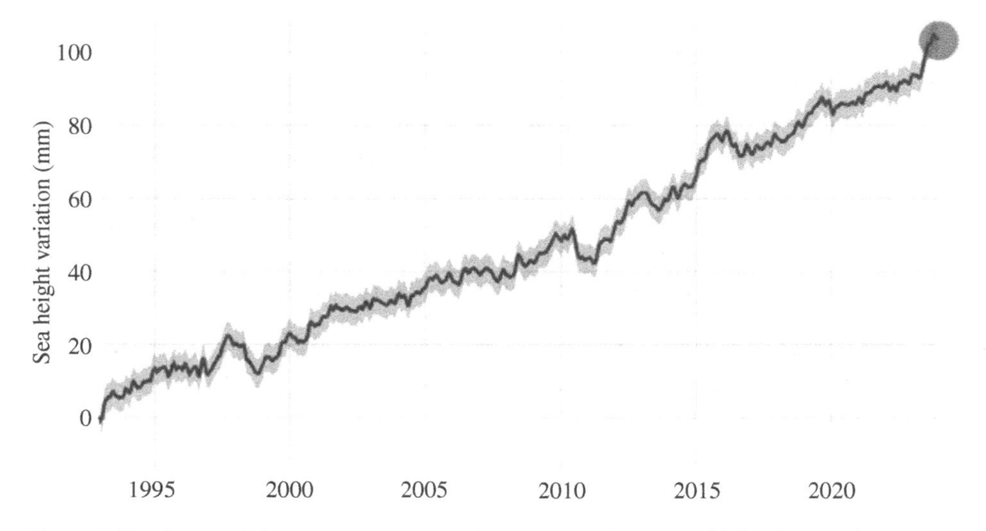

Figure 7.13 Sea level rise measured by satellite altimetry. *Source*: NASA/Public domain.

7.7 Sea Level Rise

7.7.1 Sea Level Rise Components

Sea Level Rise (SLR) refers to an increase in mean sea level with time because of climate change. SLR is occurring on account of global warming, with the two major causes being associated with thermal expansion caused by the warming of the oceans (since water expands as it warms) and with the loss of land-based ice (such as glaciers and polar ice caps) due to increased melting.

SLR comprises global SLR, which refers to SLR averaged across the globe and referenced in effect to the center of the earth, together with a regional adjustment that is needed since the sea level is nonuniform across the globe because of oceanographic, gravitational, and deformational factors.

In addition, because of slow uplift/subsidence of the land itself (see Section 7.7.3) and because coastal development work requires future water levels relative to local land surfaces, coastal engineering projects require that SLR needs to be adjusted to *Relative Sea Level Rise* (RSLR). This corresponds to SLR minus local land uplift, so that water elevations relative to local land surfaces are uninfluenced by local land uplift/subsidence.

7.7.2 Sea Level Rise Measurements

Measurements of sea level rise have been made using tide gauge records since about 1870, while methods of satellite altimetry (the measurement of elevation) have been used since about 1992.

Measurements indicate that the average rate of sea level rise since 1870 had been about 1.7 mm/yr, which is significantly larger than the average rate over the last several thousand years, and that the current rate is about 3.5 mm/yr. Figure 7.13, which is available from the NASA (National Aeronautics and Space Administration) Sea Level website via the link below,[10] shows the measured increase in global SLR since satellite measurements commenced.

7.7.3 Land Uplift/Subsidence

Land uplift/subsidence may occur because of various geophysical and hydrological phenomena, including plate tectonic movements, earthquakes, gravitational adjustments due to changing mass distributions, melting of permafrost, and changing groundwater patterns. In particular, glacial isostatic adjustment, also referred to as isostatic rebound, is a key contributor to land uplift. This is a gradual rise of landmasses that were depressed by the weight of ice sheets during glacial periods in parts of Northern Canada and Northern Europe. Figure 7.14 shows land uplift/subsidence for Canada based on the model NAV83v70VG described by Robin et al. (2020).

However, changes may be relatively localized. For example, uplift rates reported in 2020 for British Columbia have included −1.3 mm/yr for Vancouver (i.e. subsidence) and +3.3 mm/yr for Campbell River, which is only about 180 km away.

10 climate.nasa.gov/vital-signs/sea-level.

probability and extent of overtopping, inundation and/or wetting during a particular extreme storm, and the probability of these over the design life of a facility. Furthermore, a risk analysis that considers both the probability and the consequence of different types and extents of flooding may be undertaken.

In the above context, different jurisdictions have different requirements with respect to the determination of flood levels, some more and some less prescriptive. In the following subsections, different kinds of coastal flood levels are outlined in turn: the Flood Construction Level (FCL) used in British Columbia, the Base Flood Elevation and Design Flood Elevation adopted in the US, the dike crest elevation, and the tsunami flood level.

7.9.1 Flood Construction Level

First, the determination of the FCL as adopted in British Columbia is described below in some detail, so as to provide a comprehensive description of the kinds of considerations that may be taken into account in developing flood levels. The FCL refers to the required minimum base elevation of a building and relies on a 200-yr return period, an end-year of 2100 and prescribed freeboard levels. However, its use may be adapted to other applications (e.g. other kinds of infrastructure and land usage) through modifications to the bases by which the various terms are prescribed, for example, with respect to the return period, end-year selection, runup definition, and freeboard allowance. The outline given below relates to this broader context.

The FCL is obtained as

$$\text{FCL} = \text{Tide Level} + \text{Storm Surge} + \text{Relative Sea Level Rise} + \text{Wave Runup} + \text{Freeboard}$$

This addition is illustrated schematically in Figure 7.15, which shows these components leading to the design MWL, which comprises the tide level plus storm surge plus RSLR, and the FCL itself, which comprises the MWL plus wave runup plus freeboard.

7.9.1.1 Methodology

There are two alternative approaches by which some FCL components may be combined, referred to as the combined method and the probabilistic method.

Combined method. In the combined method, the high tide level is taken as Higher High Water Large Tide (HHWLT), and the storm surge and wave runup are each obtained for a storm with a 200-yr return

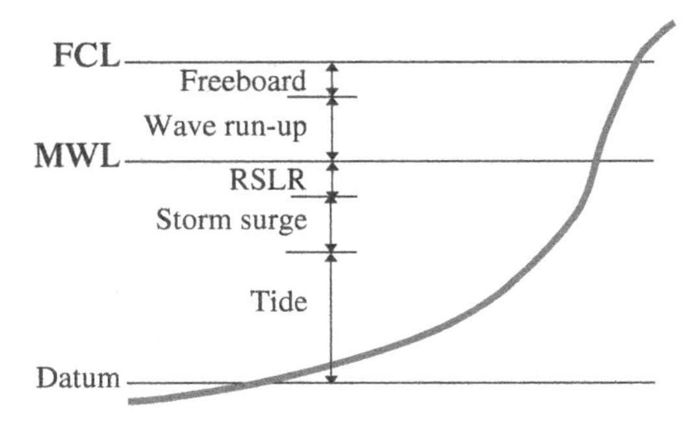

Figure 7.15 Depiction of flood construction level components.

period, referred to as a designated storm. The FCL is then obtained as the sum of the above three components, plus RSLR, plus a freeboard of 0.3 m. Since the method makes the conservative assumption that the designated storm occurs in conjunction with HHWLT, the freeboard is lower than the 0.6 m value used for the probabilistic method. The combined method is relatively straightforward to apply when the storm surge is estimated independently of the tide level.

Probabilistic method. The probabilistic method instead takes account of the joint probability of occurrence of high tide levels and large storm surge by determining the 200-yr water level due to these two components taken together. The FCL is then obtained as this combined tide and storm surge level, plus the wave runup in the designated storm, plus RSLR, plus a freeboard of 0.6 m. The probabilistic method is straightforward to apply when utilizing tide measurements, since such measurements incorporate both tides and storm surge.

7.9.1.2 Tide Level and Storm Surge

For the combined method, the 200-yr storm surge may be obtained by referring to available reports, from numerical or desktop modeling, or by undertaking an EVA of storm surge data from tide station records (i.e. using measured minus predicted tide levels). For the probabilistic method, the 200-yr tide plus storm surge is usually obtained by undertaking an EVA of tide measurements, since these include both tide and storm surge. Wave setup may be included either within storm surge or within wave runup, as may be relevant, but not both.

7.9.1.3 Relative Sea Level Rise

The RSLR is obtained for a specified time horizon, usually taken as the end-year of a project's design life or as the year 2100, relative to the current time. A modern approach to determining the corresponding RSLR value was described in Section 7.7.

7.9.1.4 Wave Runup

The 200-yr wave runup is obtained through a coastal engineering analysis that typically takes account of available wind data, wave hindcasting, and wave transformations to the nearshore, with the eventual wave runup then dependent also on the local seabed profile, wave breaking, and the characteristics of the shoreline structure. In some studies, some or all of these steps may be combined through a suitable numerical model, while in others they may be accomplished through spreadsheet calculations.

Typically, the wave runup is expressed either as $\overline{R}$, which is the mean runup of all the waves in an extreme storm, or as $R_{2\%}$, which is the runup value that is exceeded by 2% of the waves in an extreme storm. The former is relevant with respect to elevations that may safely encounter limited inundation from higher waves during an extreme storm (e.g. a parking area), whereas the latter is relevant with respect to elevations that should only encounter very occasional or no overtopping or wetting during an extreme storm (e.g. the floor level of a building). The Coastal Engineering Manual (2002) and the EurOtop Manual (2018) contain information on estimating wave runup for various shoreline profiles and shoreline structural configurations.

7.9.2 Base Flood and Design Flood Elevations

Analogous to the FCL, coastal flooding levels in the US are usually determined with respect to the Base Flood Elevation, under the jurisdiction of FEMA. These are shown on "Flood Insurance Rate Maps," which are now provided in digital form and so referred to as "Digital Flood Insurance Rate Maps." The Base Flood Elevation corresponds to the flood level resulting from a flood with an AEP of 1%. Its determination is summarized in FEMA (2011). The calculation procedures are similar to those described with respect to the FCL, but with a few key differences. Specifically, the Base Flood Elevation is based on a 1% AEP rather than on a 0.5% AEP and it does not include freeboard. Also, the wave setup and wave runup are determined separately, so that wave setup contributes to the MWL that is used to determine wave runup. Furthermore, it is not determined independently but is obtained by accessing a relevant FEMA portal or site (depending on location) with respect to a specific building footprint.

Beyond the Base Flood Elevation, local jurisdictions and design codes may require conformance to a Design Flood Elevation that is the same or higher, based on selecting a flood level AEP of less than 1% and/or the inclusion of a freeboard.

7.9.3 Dike Crest Elevation

The preceding procedure for estimating the FCL may also be applied to a determination of the crest elevation of coastal dikes. However, as an alternative, the dike crest elevation may also be calculated so as to take account of the extent of overtopping during the designated storm that is considered acceptable. The general procedure would then be to establish suitable criteria for tolerable limits to wave overtopping, for example, by taking account of the size and use of the receiving area, the extent and magnitude of drainage ditches and pumping systems, existing and anticipated land uses, and pedestrian and vehicle access (as well as potential closures during extreme events). The dike crest elevation above the MWL is then estimated such that the extent of overtopping occurring during the designated storm is matched to the level that is considered tolerable. Section 8.3.3 provides additional information on levels of overtopping considered tolerable and on the estimation of overtopping rates.

7.9.4 Tsunami Flood Level

Analogous to the case of storm-related flooding, the maximum water level due to tsunami flooding may be developed as

$$\text{Tsunami Flood Level} = \text{High Tide Level} + \text{Relative Sea Level Rise} + \text{Tsunami Runup} + \text{Freeboard}$$

There are two features of tsunamis that especially distinguish them from storms in influencing the tsunami flood level. First, since a tsunami is a rare event (e.g. once in a few hundred years to once in over a thousand years), there is insufficient information to develop runup values corresponding to a specified AEP, as in the case of storms. Instead, a different approach is adopted, relying on a consideration of different past or potential tsunami sources or events, along with an assessment of the resulting runup values at a site. Such an assessment would normally be based on a numerical modeling study or by making reference to previous studies and reports.

Second, because tsunamis are rare events, their simultaneous occurrence with an extreme storm is discounted. Furthermore, the probability of a tsunami occurring simultaneously with the highest annual tide levels is much more remote than the tsunami event itself. Even so, it is common to consider the tsunami to occur in combination with MHHW (i.e. HHWMT in Canada). However, a probabilistic and/or risk analysis may be undertaken so as to incorporate instead a lower tide level.

Example 7.3 ***RSLR and Design Flood Level***

On the basis of NASA's IPCC AR6 Sea Level Rise Projection Tool, what is the mean RSLR for 2080 relative to 2025, and the RSLR value for 2080 relative to 2025 as needed for flood elevation calculations, both for a coastal site near Port Townsend, WA. For this location, the 200-yr combined tide plus storm surge elevation (including wave setup) is 3.3 m (MLLW datum) and the 200-yr wave runup (2% exceedance value) is 1.45 m. What should be the crest elevation of a seawall if it is designed on the basis of a 200-yr storm event, a freeboard of 0.3 m, and a design life extending to 2080?

Solution

RSLR

NASA's Sea Level Rise projection tool provides 5%, 50%, and 95% exceedance RSLR values denoted $S_{5\%}$, $S_{50\%}$, and $S_{95\%}$, for RCP8.5 for Port Townsend as follows:

	$S_{5\%}$	$S_{50\%}$	$S_{95\%}$
2020:	0.000 m	0.040 m	0.086 m
2030:	0.033 m	0.070 m	0.132 m
2080:	0.252 m	0.401 m	0.747 m

By interpolation, $S_{50\%}$ for 2025 = 0.055 m
Mean RSLR increase = 0.401 − 0.055 = 0.346 m
For 2080, $S_{95\%} - S_{5\%} = 0.747 - 0.252 = 0.495$ m
Required RSLR = $0.346 + 0.2 \times 0.495 = 0.445$ m

Seawall Crest Elevation

Tide + storm surge: 3.30 m MLLW
RSLR: 0.45 m
Wave runup: 1.45 m
Freeboard: 0.30 m
Seawall crest elevation: 5.50 m MLLW

■➔

7.9.5 Probability of Coastal Flooding

While the preceding guidelines establish design flood levels, it is of interest to estimate the probability that coastal flooding occurs over the life of a project. This may be taken to occur when the water elevation, suitably defined and denoted here as X, is taken to exceed a structure crest elevation h. However, since flooding or overtopping within a storm may occur to different degrees of severity, alternative definitions of X may be adopted. One such approach is to consider three levels of potential flooding. Sustained inundation is taken to occur when X is defined as equal to the MWL during a storm, so that $X = h$ corresponds to continuous submergence of the area landward of the structure crest over the course of an extreme storm. Intermittent inundation is taken to occur when $X = \text{MWL} + \overline{R}$, where $\overline{R}$ is the mean wave runup, so that $X = \text{h}$ then corresponds to about half the waves overtopping a structure crest during the storm. Finally, wetting is taken to occur when X is defined as equal to $\text{MWL} + R_{2\%}$, where $R_{2\%}$ denotes the wave runup that is exceeded by 2% of the waves in the storm, so that $X = h$ then corresponds to only occasional overtopping during the storm. Thus, more risk-tolerant infrastructure may rely on the first of these, while less risk-tolerant infrastructure may rely on the third of these.

Regardless of this choice, the probability of any one of these occurring within the project life L may be estimated in the following manner. First, the AEP of X, corresponding to $Q(X)$, is determined on the basis of an estimated extreme probability distribution of X, such as the Gumbel distribution. The AEP of overtopping then corresponds to the value of $Q(X)$ when X is set to h, i.e. $Q(h)$. For example, when the dike elevation h is set to the 200-yr value of X with no freeboard allowance, then the AEP of flooding corresponds to $Q(h) = 0.05$, whereas the inclusion of some freeboard allowance corresponds to $Q(h) < 0.05$. In the absence of RSLR, the encounter probability E of $X > h$ occurring over the project life L is then given simply as $E = 1 - \exp[-Q(h)\, L]$.

Because of the continual increase in RSLR each year, the probability of overtopping each year increases from a minimum value in the initial years of a project to a maximum value in the final years of the project. Specifically, the AEP of overtopping in the nth year, denoted Q_n, corresponds to $Q(h - R_n)$, where R_n denotes the RSLR value in the nth year. That is, the increased Q_n value may be considered to correspond to the dike elevation being in effect lowered by an amount R_n.

The probability of overtopping E over a duration of N years may then be estimated from the following equation:

$$E = 1 - \prod_{n=1}^{n=N} (1 - Q_n)$$

where Π is the product operator and n represents successive years, with $n = 1$ corresponding to the initial year of the project. One of the problems in this chapter enables the reader to explore this approach in relation to a particular case.

←■

7.9.6 Consequences of Coastal Flooding

Beyond the establishment of design flood levels as described in this section, the consequences of overtopping of dikes and other coastal defenses also require consideration. Thus, attention is now paid to wave propagation overland, required setbacks for buildings, flood protection systems, temporary flood barriers, and the floodproofing of buildings.

Overland wave propagation. When coastal flooding occurs, the attenuation of waves as they propagate overland is relevant to determining design flood levels at a site set back from the shoreline. This situation has been described by FEMA (2023). The degree of wave attenuation will depend on breaking limits on the wave crest elevation associated with lower depths overland, energy dissipation due to wave breaking and bottom friction (possibly influenced by the kinds and extent of vegetation encountered), bottom percolation (if relevant), and the effects of buildings and other obstacles on the waves. When overland wave propagation occurs over larger areas, associated with extensive inundation due to storm surge, a computer model may be used to describe overland wave propagation taking account of the features described above. When flooding occurs over a limited width of shoreline, wave attenuation may also be associated with wave directional spreading. In addition, when the MWL remains below the ground level, then only the crests of larger individual waves overtop and propagate overland. In such a case, the preceding considerations continue to apply but require modification, and as well the individual waves take the form of discrete, unstable water masses that also attenuate by broadening in the direction of wave travel.

Setbacks. In some jurisdictions, setbacks are established in order to ensure that any development occurs sufficiently far away from areas of potential erosion or flooding. For example, a building setback may be specified so as to correspond to a given distance landward of a suitably defined boundary. The building setback is intended to take account of all aspects of a flood hazard associated with future water levels, including potential wave, debris, and splash impacts on buildings. Other circumstances requiring setbacks relate to a tsunami hazard and to lots located above coastal bluffs that are susceptible to toe erosion from storms.

Flood protection systems. In the event of flooding over a dike, a flood protection system is used to remove flood waters as completely and as quickly as possible. Apart from the dike itself, a flood protection system includes a number of components. Thus, a pump station acting in conjunction with a flood box transports pumped floodwater through a culvert past the dike to the sea, with a flap-gate mounted at the discharge end so as to prevent any backflow from the sea. Additionally, a drainage system comprised of culverts, drainage canals, ditches, drains, and/or storage lagoons is intended to limit floodwaters and convey them back to the sea. Finally, floodgates are built at the discharge end of storm sewers so as to prevent backflow from the sea or a river.

Temporary flood barriers. Temporary flood barriers may be secured across gaps or around a perimeter so as to prevent flooding with respect to infrequent, severe high-water events. These may entail various devices, such as a system of floodwall panels, elongated flexible tubes filled with water, and specially designed sandbags. A key requirement is that such devices need to be deployed at short notice.

Floodproofing of buildings. For buildings in a flood-prone area, the design, construction, and operation of a building can incorporate various aspects of floodproofing. ASCE (2006) provides a standard for the design and construction of such buildings. This includes basic siting, design, and construction requirements for buildings and floodproofing measures that include the following. Dry floodproofing refers to perimeter flood barriers, temporary floodwalls, flood barrier panels, flood-proof windows and doors, and flood pump systems that are intended to prevent water from entering a building. Wet floodproofing refers to measures that limit flood damage, recognizing that wetting to interior spaces may occur. Approaches include fitting a building with wall openings or "flood vents" to enable flood water to enter and exit certain enclosed spaces, such as crawl spaces, basements, and garages, so as to prevent hydrostatic pressure buildup that can destroy walls or foundations. Other measures include the use of flood-resistant building materials and waterproof coatings to minimize water damage. Access considerations relate, for example, to living areas that are located at a sufficiently high elevation and to a basement design with two staircases to provide occupants with a safe means of egress. The design and placement of utilities and equipment, including electrical, plumbing, and mechanical/HVAC systems, may include the need to place some equipment at a sufficiently high elevation and the design of plumbing systems that incorporate a backflow prevention system.

7.10 Coastal Currents

This section provides a general summary of different kinds of coastal currents.

Ocean circulation patterns. First, mention is made of large-scale ocean current patterns associated with wind patterns, water density changes (due to surface water temperature and salinity changes), and Coriolis effects. The patterns correspond to large eddies or gyres in each ocean basin – the larger ones rotating clockwise in the Northern Hemisphere and counterclockwise in the Southern Hemisphere, as indicated in Figure 7.16. A well-known example is the Gulf Stream.

Tidal currents. Tidal currents arise in response to surface elevation changes as tides flood and ebb. In the case of an enclosed bay with a narrow entrance, the maximum tidal currents at the entrance depend on the volume of water filling a bay in each tidal cycle, and the cross-sectional area of the entrance through which the water must pass. Numerical models of tidal currents are available, based on tidal fluctuations and local bathymetry, taking account of friction at the seabed. Tidal data sources may provide information on tidal currents – see the link below[13] to NOAA's Tides and Currents website for the US and the link

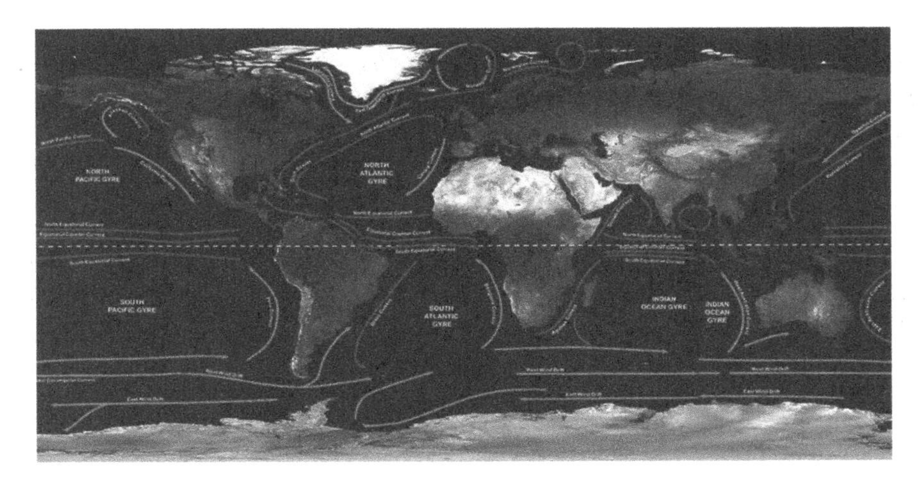

Figure 7.16 Gyres in ocean basins worldwide. *Source*: NOAA/Public domain.

13 tidesandcurrents.noaa.gov.

below[14] to DFO's Tides, Currents, and Water Levels website for Canada. However, it must be borne in mind that these currents typically vary significantly with location.

Wind-induced currents. These are due to winds blowing over the ocean surface. Different design codes have provided information on the magnitude and profile of the resulting current in terms of the wind speed. In one such example, the current at the water surface is taken as 3% of the wind speed at an elevation of 10 m, and this is taken to decay linearly to zero at 50 m below the surface.

Mass-transport velocity. This refers to second-order effects in a propagating wave train, such that the water particle trajectories are not closed orbits but include a steady drifting motion. Predictions of the mass-transport velocity and its distribution with elevation are available. However, this is not usually a consideration in an assessment of currents to be relied upon in a coastal engineering project.

Longshore current. This refers to a current parallel to the shore induced by waves approaching a shoreline obliquely. This current is important in causing sediment movement parallel to a beach, termed littoral drift, and is considered in Section 9.6.2.

Rip currents. These correspond to a strong outflow away from a shoreline acting over a narrow width at certain locations along a beach. They may occur when there is a pronounced variability in the wave breaking pattern in the longshore direction or at local geometric or bathymetric irregularities. These currents are dangerous to swimmers and may cause localized sediment erosion.

River currents. Finally, as a river discharges into a sea, the river current extends into sea, with the current profile influenced by tidal behavior, mixing processes, and salinity variations.

Problems

7.1 Using the relevant NOAA website, what were the elevations and local times of the higher high and lower low tide levels above the datum MLLW at Point Roberts Marina, WA, on December 3, 2024? For that day, plot the surface elevation vs. time (in December 3 hours). Estimate the maximum rate of change of the surface elevation that day, and the range of times that day that this rate was near a maximum. Based on continuity considerations (i.e. by equating the increase or decrease in volume in the marina per unit time to the volume flow rate into or out of the marina), develop an equation for the current speed U in the marina entrance channel in terms of the marina surface area S, the entrance channel width w, the entrance channel depth d, and the rate of change of tide level in the marina. If $S = 145\,000\ \text{m}^2$, $w = 48$ m, and the seabed is 4.2 m below MLLW in the entrance channel, estimate the maximum tide-induced current in the entrance channel that day.

7.2 A pressure sensor on the seabed at an offshore location is used as part of a tsunami warning system for a community located at the adjacent shoreline. For the seabed profile indicated in the sketch below, estimate the duration for a tsunami to travel from the offshore location A to the shoreline location B. Use a numerical scheme in which the sloping portion of the profile is discretized into

14 tides.gc.ca.

20 segments. What would this duration be if instead the varying depth of the sloping portion of the profile is approximated by its average depth?

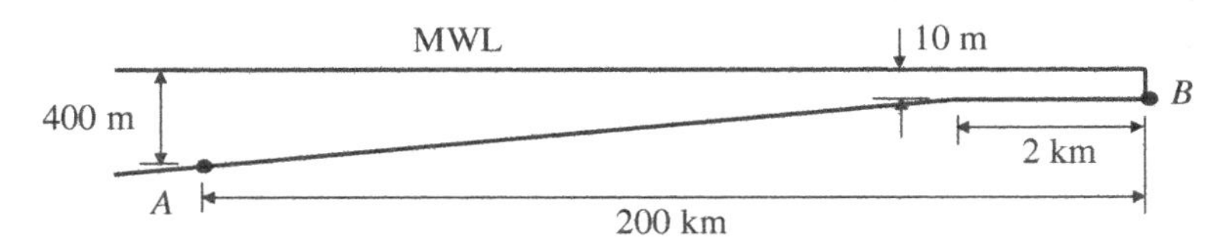

7.3 The seabed profile adjacent to a site corresponds to a shelf of uniform slope extending from deep water offshore to the shoreline. For the seabed profile indicated in the sketch below, estimate the wind setup at the shore for a design wind of 100 km/hr that blows directly onshore. Assume that the effective drag coefficient $\kappa = 3.3 \times 10^{-6}$. Use a numerical scheme in which the shelf is discretized into 20 segments. What would this setup be if instead the varying depth of the shelf is approximated by a uniform depth that is 0.25 times its maximum depth?

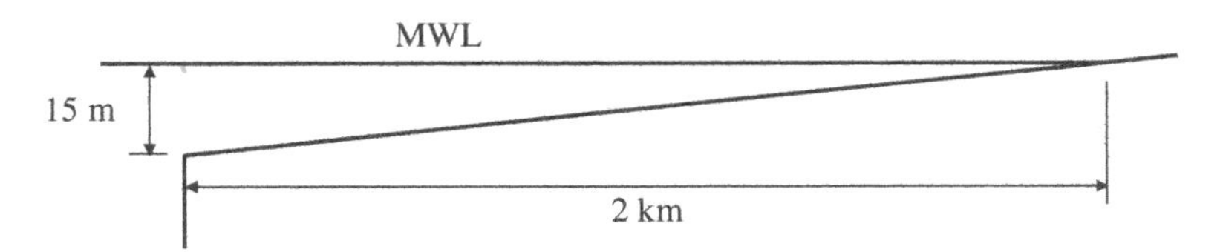

7.4 The text's companion site allows for a download of the spreadsheet *Problem Data*. The tab "Problem 7.4" contains data corresponding to hourly measurements of tidal elevations relative to CD at Vancouver, BC, over the period January 1, 1950 to December 31, 2022. Use the data file to estimate the tide plus storm surge elevation with return periods of 100 and 500 yr at this location.

7.5 *Note: alternate versions of this problem reflecting US practice and Canadian practice are provided. Both versions rely on the use of online portals as may be relevant.*

(a) ***US practice***: An estimate of the Design Flood Elevation for the year 2100 at a site near Friday Harbor, WA, is required on the basis of an AEP of 0.5% and a freeboard of 0.3 m. The tide plus storm surge elevation is taken as the highest observed tide as reported on the relevant NOAA portal, the wave runup (2% exceedance value) is 1.3 m, and the wave setup is negligible. What is the RSLR to be used in the determination of the Design Flood Elevation? What are the Design Flood Elevation values relative to MLLW and relative to NAVD 88?

(b) ***Canadian practice***: The FCL for the year 2100 at a site near Point Atkinson, BC, is required. The 200-yr storm surge is 1.0 m and the 200-yr wave runup (2% exceedance value) is 1.4 m. What is the RSLR to be used in the determination of the FCL? What are the FCL values relative to CD and relative to CGVD2013?

7.6 ■ Coastal flood water levels, denoted X, which take account of tide, storm surge, and wave runup, are defined on the basis of the average runup, corresponding to intermittent inundation during a storm. The 50- and 200-yr values of X at a site are 5.7 and 6.4 m, respectively. Assuming a Gumbel distribution for X, what are the AEP values for intermittent inundation if the dike crest elevation is set at 6.4, 7.0, and 7.3 m in turn? Excluding considerations of RSLR, what will be the corresponding encounter probabilities for a 50-yr design life? For a dike crest elevation of 7.0 m, what will be the modified encounter probability of intermittent inundation during the design life if RSLR increases uniformly by 0.5 m over that period?

8

Coastal Structures

8.1 Introduction

This chapter describes the effects of waves on coastal structures, with a focus on methods to determine wave loads on coastal structures, wave runup and overtopping of shoreline structures, the stability of rubble-mound structures, and the motions of floating structures. The chapter also includes a consideration of other forms of loading on coastal structures. Finally, as supplementary information, it includes an outline of renewable energy extraction from ocean sources. Distinct from this chapter's coverage, Chapter 11 focuses on coastal infrastructure design, including an identification of relevant design criteria and the application of selected design tools.

8.1.1 Categories of Structure

Coastal structures may be categorized in various ways. One approach is based on their functional requirements, while another is based on the nature of their interactions with waves. From the viewpoint of functional requirements, categories of structure include the following: shoreline protection structures that include different kinds of seawalls and rubble-mound structures that limit backshore erosion, overtopping and flooding; flood control infrastructure that includes dikes, levees, flood walls, and flood barriers with retractable flood gates; sediment control structures that include groins and offshore breakwaters that limit sediment movement; breakwaters that include fixed and floating breakwaters and that provide protection from waves; pipelines and outfalls; and harbor and port structures that include berths, docks, piled wharfs, and dolphins.

With respect to this chapter's organization, it is more effective to categorize structures from the viewpoint of their interactions with waves, especially with respect to wave loads and/or wave runup. Broadly, these include seawalls, rubble-mound structures in the form of either shoreline protection or breakwaters, slender-member structures, large structures, and floating structures. Examples of these are illustrated in Figure 8.1.

Summary descriptions of these kinds of coastal structures are given below.

Seawalls. Seawalls and other impermeable shoreline structures largely reflect the incident waves, while encountering wave runup and possibly wave overtopping. Figure 8.1a shows one such example. As an extension to this category of structure, an impermeable wall-type structure may form a breakwater with water on both the exposed and sheltered sides, referred to as a caisson breakwater.

An Introduction to Coastal Engineering, First Edition. Michael Isaacson.

Companion website: www.wiley.com/go/coastalengineering

Figure 8.1 Examples of categories of coastal structures. (a) Seawall – Vancouver, BC, (b) rubble-mound shoreline protection – Vancouver, BC, (c) rubble-mound breakwater – Gibsons, BC, (d) piled structure – Gibsons, BC, (e) large structure – Rio-Antirrio Bridge, Greece, (f) floating structure – Horseshoe Bay, BC. *Sources*: (c) ShoreZone/CC BY 3.0. (e) Rio-Antirrio Ferries/Wikimedia Commons/CC BY 3.0.

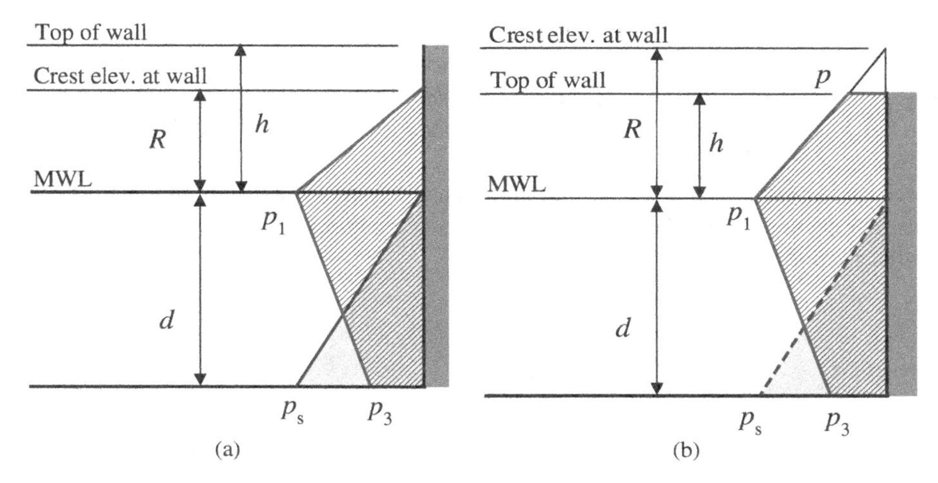

Figure 8.5 Definition sketch for the Goda formulation. (a) No overtopping ($h > R$), (b) overtopping ($h < R$).

locations that lead to the loads as follows:

$$R = 0.75(1 + \cos\beta)H_\mathrm{d}$$

$$p_1 = 0.5(1 + \cos\beta)(\alpha_1 + \alpha_2\cos^2\beta)\rho g H_\mathrm{d}$$

$$p_2 = \begin{cases} (1 - \mathrm{h}/R)\,p_1 & \text{for} \quad R > h \\ 0 & \text{for} \quad R \leq h \end{cases}$$

$$p_3 = \alpha_3 p_1$$

$$p_\mathrm{s} = \rho g d$$

In the above, H_d is the design wave height, β is the angle between the incident wave crests and the face of the wall, and d_s is the water depth at a distance $5H_\mathrm{s}$ in front of the wall. Also,

$$\alpha_1 = 0.6 + 0.5\left[\frac{2kd_\mathrm{s}}{\sinh(2kd_\mathrm{s})}\right]^2$$

$$\alpha_2 = \min\begin{cases} \left(\dfrac{d_\mathrm{s} - d}{3d_\mathrm{s}}\right)\left(\dfrac{H_\mathrm{d}}{d}\right)^2 \\ \dfrac{2d}{H_\mathrm{d}} \end{cases}$$

$$\alpha_3 = 1 - \frac{d}{d_\mathrm{s}}\left[1 - \frac{1}{\cosh(kd_\mathrm{s})}\right]$$

The design height H_d is taken as $1.8H_\mathrm{s}$ if the structure is outside the surf zone or is otherwise taken as the maximum breaking wave height at a distance $5H_\mathrm{s}$ in front of the wall.

In an extension to Goda's initial formulation, intended to account for breaking waves, α_2 is replaced by a different parameter α_I if $\alpha_I > \alpha_2$, where α_I may be obtained by a formula given in the Coastal Engineering Manual (2002). The parameter α_I, which leads to a greater load associated with waves breaking onto the wall, depends on the berm width, which is the horizontal distance from the wall to where the seabed begins to slope downward.

The maximum force and overturning moment may be obtained from suitable integrations of the pressure distributions. The sectional force F'_s and sectional overturning moment M'_s associated with the hydrostatic pressure distribution are thereby given as

$$F'_s = \frac{1}{2} p_s d$$

$$M'_s = \frac{1}{6} p_s d^2$$

The sectional force F'_d and sectional overturning moment M'_d associated with the hydrodynamic pressure distribution are given by

$$F'_d = \frac{1}{2} p_1 R + \frac{1}{2}(p_1 + p_3)d - \frac{1}{2} p_2 (R - h)$$

$$M'_d = \frac{1}{2} p_3 d^2 + \frac{1}{3}(p_1 - p_3)d^2 + \frac{1}{2} p_1 R \left(\frac{R}{3} + d\right) - \frac{1}{2} p_2 (R - h)\left[\frac{R - h}{3} + h + d\right]$$

The total sectional force is then given as $F' = F'_s + F'_d$ and the total sectional overturning moment is given as $M' = M'_s + M'_d$. The above formulation applies equally to the overtopping ($R > h$) and no overtopping ($R \leq h$) cases, bearing in mind that $p_2 = 0$ in the latter case. As before, if the wall functions as a breakwater with water on both sides, then, assuming still water on the sheltered side, the hydrostatic load components acting on both sides of the breakwater cancel out.

Reference Solution A8 in Appendix A includes a spreadsheet solution based on the Goda formulation, taking account of the extension relating to α_I as indicated above.

8.2.5 Related Impermeable Structures

The preceding considerations for a vertical wall may be extended to a sloping, impermeable wall that may or may not have a curved profile. Descriptions of the runup and wave loads on sloping walls are provided in various design guides, notably the Coastal Engineering Manual (2002). While some information on wave runup estimates on vertical walls is included above, a more general description of wave runup and overtopping for sloping impermeable walls is deferred to Section 8.3.3, where the case of rubble-mound structures is also considered.

Example 8.1 ***Loads on a Seawall***

A regular wave train with $H = 0.8$ m and $L = 24$ m approaches a vertical seawall normally in water of constant depth $d = 4$ m. Based on the Miche-Rundgren/Sainflou methods, what are the wave runup, the maximum pressure on the wall at the seabed, and the maximum force per unit width on the wall?

Solution

Specified parameters

$g = 9.80665\ \text{m/s}^2$
$H = 0.8\ \text{m}$
$L = 24.0\ \text{m}$
$d = 4.0\ \text{m}$

Wave runup

$\Delta = \pi H^2/[L \tanh(kd)] = 0.1\ \text{m}$
$R = H + \Delta = 0.9\ \text{m}$

Maximum pressure and force

$p_o = \rho g[d + H/\cosh(kd)] = 45.2\ \text{kN/m}^2$
$F'_m = \frac{1}{2}\, p_o\, (d + R) = 111.0\ \text{kN/m}$

8.3 Rubble-Mound Structures

Rubble-mound structures are composed of individual rocks or artificial armor units as indicated in Figure 8.6 and may either be placed against a shoreline as a form of shoreline protection (Figure 8.6a) or serve as a breakwater with water on both sides (Figure 8.6b). *Riprap* is an alternate, more informal

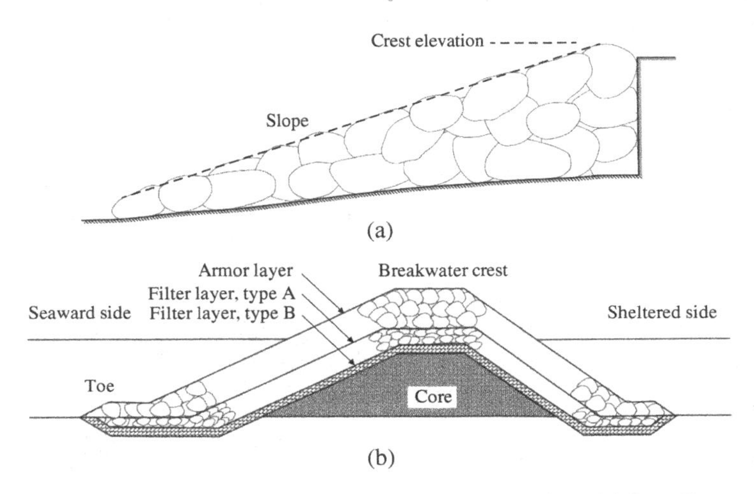

Figure 8.6 Examples of rubble-mound structure sections. (a) Shoreline protection section with a single layer of riprap, (b) breakwater section with multiple layers.

term that is also used. Rubble-mound structures may range from a simple protection scheme requiring modest specifications and a homogeneous layer of rock, as illustrated in Figure 8.6a, to a comprehensive breakwater section as illustrated in Figure 8.6b. The latter may include distinctly specified armor layers (which may include specially shaped artificial units – not shown), distinct filter layers, a breakwater core, specially designed toe protection, and different designs on the seaward and sheltered sides.

A primary consideration in the design of such structures is armor stability or the extent of movement of these units under storm conditions. Other key considerations include wave runup and overtopping, to be considered in Section 8.3.3.

Armor units consist of quarry stone that may be more rounded or angular or they may be precast concrete artificial units designed to increase stability by interlocking. A wide variety of artificial units have been developed and deployed. Examples of particular shapes include those referred to as Accropode, Cube, Dolos, Stabit, Tetrapod, and Xbloc, with the shapes of some units modified in subsequent versions. Figure 8.7 shows an example of a natural rock breakwater (Figure 8.7a) and three examples of free-standing armor units – Tetrapod (Figure 8.7b), Dolos (Figure 8.7c), and Xbloc (Figure 8.7d) units.

Figure 8.7 Examples of a rock breakwater and free-standing armor units. (a) Rock breakwater, (b) Tetrapod units, (c) Dolos units, (d) Xbloc units. *Sources*: (a) Ian Capper/CC BY SA 2.0. (b) Pjotr Mahhonin/Wikimedia Commons/CC BY SA 4.0. (c) Fernando Losada Rodríguez/Wikimedia Commons/CC BY SA 4.0. (d) Oliver Dixon/CC BY SA 2.0.

The selection of a particular kind of armor unit and its size depends on unit stability under storm conditions. In addition, design considerations for relatively large units include structural strength (including possible reinforcing), durability, material properties, and methods of fabrication, transportation, and placement.

8.3.1 Predictions of Armor Stability

A key aspect of the design of a rubble-mound structure is the ability to predict conditions under which armor units remain stable, since failure is often triggered by an excessive movement of units. Furthermore, once units are dislodged, an indication of the nature and extent of subsequent damage is also required.

Alternate approaches to predicting armor stability are based on the Hudson equation, first proposed in 1958, and the Van der Meer equations, first proposed in 1988. The former is simpler to use, while the latter entails additional parameters and has been extended in various ways. These are summarized in turn.

8.3.1.1 Hudson Equation

This equation provides the median mass M of individual units as follows:

$$M = \frac{\rho_s H^3}{K_D\, s'^3 \cot\beta}$$

where ρ_s is the density of the unit material, often taken as 2650 kg/m^3 for rock and 2400 kg/m^3 for concrete units. H is the incident wave height, K_D is an empirical stability coefficient, referred to as an armor stability coefficient, $s' = \rho_s/\rho - 1$ is the submerged specific gravity of the units, and $\tan\beta$ is the breakwater slope. In using the Hudson equation, H is taken as $H_{1/10} = 1.27\, H_s$, where H_s is the significant wave height near the structure.

The Hudson equation may also be expressed in an alternative form so as to provide the *nominal diameter D*, which is defined as the side length of a cube of the same mass. The relationship between nominal diameter D and the unit's mass M is $D = (M/\rho_s)^{1/3}$. The equation now takes the form:

$$\frac{H}{s'D} = (K_D \cot\beta)^{1/3}$$

The Shore Protection Manual (1984) provides K_D values corresponding to no damage or limited damage for various armor units, different placements, different locations (breakwater head or trunk), and the number of layers and for breaking or nonbreaking wave conditions. These include the following values of K_D for randomly placed units in two layers with respect to a breakwater trunk:

$$K_D = \begin{cases} 1.2 & \text{for smooth, rounded rock subjected to breaking waves} \\ 2.4 & \text{for smooth, rounded rock subjected to nonbreaking waves} \\ 2.0 & \text{for rough, angular rock subjected to breaking waves} \\ 4.0 & \text{for rough, angular rock subjected to nonbreaking waves} \end{cases}$$

K_D values are lower with respect to a breakwater head and are higher with respect to units forming additional layers. K_D values are also higher for special placements of the units and for specially designed units.

8.3.1.2 Van der Meer Equations

The original Van der Meer equations provide the nominal diameter of individual units as follows:

$$\frac{H_s}{s'D} = \begin{cases} 6.2\,P^{0.18}\left(\dfrac{S}{\sqrt{N}}\right)^{0.2} \xi^{-0.5} & \text{for} \quad \xi < \xi_c \quad \text{(plunging breakers)} \\ 1.0\,P^{-0.13}\left(\dfrac{S}{\sqrt{N}}\right)^{0.2} \sqrt{\cot\beta}\,\xi^{P} & \text{for} \quad \xi > \xi_c \quad \text{(surging breakers)} \end{cases}$$

where N is the number of waves, $S = A_e/D^2$ is termed the damage level such that $S \simeq 2$ represents an initial level of damage, and A_e is the average eroded area (near the SWL) of a cross-sectional profile. P is a notional permeability factor, with $P = 0.1$ for a relatively impermeable breakwater (e.g. two armor layers with a thin filter layer on an impermeable core), $P = 0.5$ for one armor layer on a permeable core, and $P = 0.6$ for a homogenous breakwater with no core. Also,

$$\xi = \frac{\tan\beta}{\sqrt{2\pi H/gT_z^2}}$$

$$\xi_c = [6.2\,P^{0.31}\sqrt{\tan\beta}\,]^{1/(P+0.5)}$$

where T_z is the zero-crossing period. Recall that, for the Pierson–Moskowitz spectrum, T_z is related to the peak period T_p via $T_z = 0.71T_p$. The extent of the angularity of the rock units may be taken into account by adjusting the factors 6.2 and 1.0 above, while related formulae have been provided for specially designed armor units.

8.3.1.3 Damage Progression

Beyond the prediction of armor stability, there is interest in describing the nature of the damage that occurs, quantifying the extent of damage, and considering the progression of damage that ultimately results in the failure of a rubble-mound structure.

The Shore Protection Manual (1984) refers to the percentage damage corresponding to $A_e/A_o \times 100\%$, where A_e is the average eroded area with respect to a cross-sectional profile and A_o is the area of the original profile. The Shore Protection Manual (1984) provides information on the extent of the damage that may be encountered for waves of different heights relative to the design wave height, for different kinds of armor units, placements, and locations (breakwater head or trunk), and for breaking or nonbreaking wave conditions.

The damage parameter S appearing in the van der Meer equations represents an alternative approach to quantifying the extent of damage. Thus, $S = 2$ is commonly taken as indicative of the start of damage, while increasing values of S correspond to greater levels of damage, including intermediate damage and eventually failure, with S then typically in the range 8–12. Thus, the van der Meer equations may be viewed as corresponding to a damage progression model in that they incorporate S and may be used as a basis for considering progressive levels of damage.

The Coastal Engineering Manual (2002) contains extensive information relating to armor stability and damage progression of rubble-mound structures.

Example 8.2 ***Stability of Rubble-Mound Structures***

A rubble-mound breakwater with a 1/1.5 slope is to be located at a site where the design wave corresponds to $H_s = 1.5$ m and $T_p = 5.0$ s. Using the van der Meer equations, estimate the median mass and nominal diameter of rock armor units required for no damage to the breakwater. Assume that the rock has a density $\rho_s = 2650$ kg/m, the permeability factor $P = 0.5$, and the storm duration $\tau = 6$ hr.

Solution

Specified parameters

$g = 9.80665\ \text{m/s}^2$
$\rho = 1025\ \text{kg/m}^3$
$\rho_s = 2650\ \text{kg/m}^3$
$H_s = 1.5$ m
$T_p = 5.0$ s
$m = 1/1.5$
$P = 0.5$
$\tau = 6$ hr
$S = 2$ (corresponding to the limiting case of no damage)

Van der Meer Equation Parameters

$$T_z = 0.71\ T_p = 3.6\ \text{s}$$
$$s' = \rho_s/\rho - 1 = 1.59$$
$$N = \tau/T_z = 6085$$

Require ξ, ξ_c to determine the form of wave breaking:

$$\xi = \frac{\tan\beta}{\sqrt{2\pi H_s/gT_z^2}} = 2.41$$

$$\xi_c = \left[6.2\,P^{0.31}\sqrt{\tan\beta}\right]^{1/(P+0.5)} = 4.08$$

Apply Van der Meer Equations

$\xi < \xi_c$, therefore plunging breakers

The corresponding equation is

$$\frac{H_s}{s'D} = 6.2\,P^{0.18}\left(\frac{S}{\sqrt{N}}\right)^{0.2}\xi^{-0.5} = 1.69$$

Therefore:

$$D = 0.56\ \text{m}$$
$$M = \rho_s D^3 = 463\ \text{kg}$$

8.3.2 Alternate Rubble-Mound Configurations

Beyond traditional rubble-mound breakwaters, other rubble-mound configurations that have also been used include those summarized below.

Berm breakwater. A berm breakwater is one that includes a berm comprising rocks placed near or above the SWL adjacent to the breakwater crest. This acts to reduce wave runup and overtopping and to reduce damage to the armor layers. The berm usually comprises smaller rocks and is designed to reshape into a more stable profile when exposed to storms. A berm breakwater usually results in cost savings when compared to a conventional design because of the use of smaller, more readily available materials forming the berm. Different kinds of berm breakwater include those that are designed not to reshape and those whose reshaped profile is either statically stable, with no subsequent rock movement, or dynamically stable, with some rock movement along the front slope. Also, the berm may consist of homogenous rocks or of different rock classes, usually with larger rocks near the crest of the berm. Van der Meer and Sigurdarson (2017) provide extensive information on the design and construction of berm breakwaters.

Low-crested breakwater. As its name implies, a low-crested breakwater has its crest close to the SWL and so enables some wave transmission by overtopping. Because of this, it provides reduced protection from waves when compared to a traditional rubble-mound breakwater. On the other hand, its armor stability is increased so that smaller rocks can be used at a lower cost – although its sheltered side is relatively vulnerable and so requires a modified armor layer design. As with a traditional breakwater, it also acts as a sediment barrier to protect a beach against erosion.

Submerged breakwater. A submerged breakwater remains below the SWL and so is relatively acceptable aesthetically in recreational areas. While it also enables a greater degree of wave transmission, it is designed to trigger larger waves to break so as to limit transmitted wave heights. As with a low-crested breakwater, its armor stability is improved, its sheltered side requires a modified armor layer design, and it also acts as a sediment barrier. In addition, a submerged breakwater does not limit water circulation and flushing between the structure and a beach.

Composite breakwater. A composite breakwater usually refers to a caisson breakwater placed above a rubble-mound base. A potential failure mode may correspond to waves breaking near the mound and then slamming against the vertical wall. They are similar to caisson (vertical wall) breakwaters except that the water depth in front of a caisson is reduced notably because of the rubble berm.

8.3.3 Wave Runup and Overtopping

The following summary refers to both rubble-mound structures and impermeable wall-type structures.

8.3.3.1 Wave Runup

As described in Section 7.9.1, wave runup is typically expressed as $R_{2\%}$, which is the runup value that is exceeded by 2% of the waves in an extreme storm, or as $\overline{R}$, which is the mean runup of all the waves in an extreme storm. $R_{2\%}$ is relevant with respect to elevations that should only encounter very occasional overtopping during an extreme storm, whereas $\overline{R}$ is relevant with respect to elevations that may encounter a greater level of overtopping from higher waves during an extreme storm.

Wave runup estimates for vertical walls are provided in Section 8.2. The wave runup for a sloping impermeable wall depends primarily on the surf similarity parameter, the local water depth, and the wall slope and may be notably greater than for a vertical wall as waves surge up the sloping face of the wall. The Coastal Engineering Manual (2002) and the EurOtop Manual (2018) provide runup estimates for both rubble-mound structures and impermeable walls, taking into account seabed profiles, incident wave conditions, and the characteristics of the wall or rubble-mound structure.

8.3.3.2 Wave Overtopping

Wave overtopping refers to the extent to which water flows over the crest of a shoreline structure. The *overtopping rate*, sometimes referred to as the overtopping discharge and denoted q, refers to the average overtopping discharge per m length along the shoreline for a particular storm condition or sea state (in units of m^3/s per m length or L/s per m length, where 1 m^3/s per m = 1000 L/s per m). Other overtopping parameters include the overtopping volume, which refers to the volume of water that overtops the structure crest (in units of m^3 per m length) for a single wave, a series of waves or by extension a specified storm or storm duration, and the *maximum overtopping volume*, denoted V_m, which is the maximum overtopping volume for a single wave.

Overtopping occurs primarily through the flow of water directly over the structure crest, referred to as "green water," and to a minor extent through splash or spray transported above the structure crest that is referred to as "white water" and is associated with breaking waves, exacerbated by onshore winds.

The EurOtop Manual (2018) provides estimates of the extent of overtopping for both rubble-mound structures and impermeable walls. Typically, results are presented in the form of the relative overtopping rate, $q/(gH_{st}^3)^{0.5}$, as a function of relative freeboard, h/H_{st}, where H_{st} is the significant wave height near the toe of the structure, and h is the freeboard, which is the structure crest elevation above the mean water level. This kind of relationship depends on the seabed slope or profile, the structure type and configuration, and the characteristics of the incident waves.

In undertaking a design, the estimated overtopping rate that occurs needs to be compared to an overtopping rate that is considered tolerable. Thus, a *tolerable overtopping rate* and a *tolerable maximum overtopping volume* are the values of q and V_m that are considered tolerable with respect to a particular set of potential impacts or conditions. The EurOtop Manual (2018) provides information on suitable values or ranges of q and V_m that are considered tolerable with respect to impacts on pedestrians, vehicles, property, equipment and facilities, and small craft protected by a breakwater. These depend also on the magnitude of the incident waves.

8.4 Slender Structures

This category of coastal structure refers to a structure comprising one or more small diameter structural elements, e.g. a submarine pipeline, a vertical or sloping pile, a pile-supported wharf, or a more complex framework platform. For such a structure, the member diameter D is much less than the wave length L. Therefore, the structure does not diffract the waves and waves tend to continue unimpeded past the structure. However, the orbital diameter of the flow is of the same order of magnitude as D or greater and so flow separation around the structure needs to be taken into account. Wave loads for such structures have been described in detail by Sarpkaya and Isaacson (1981).

8.4.1 Development of Morison Equation

Predictions of wave forces on such structures are based on the Morison equation. In order to examine the development of this equation, the reference case of a two-dimensional oscillatory flow past a circular section is initially considered. The force formulation is developed by considering two fundamental flows relating to this case, as sketched in Figure 8.8. Figure 8.8a indicates the steady flow of a real (viscous) fluid past a circular section, leading to flow separation and vortex shedding in the wake of the structure. Figure 8.8b refers to the unsteady (accelerating) flow of an ideal (inviscid) fluid past a circular section, for which there is no flow separation.

Each of these flows has been widely studied. For the first of these (Figure 8.8a), corresponding to a steady flow of a real fluid past a circular cylinder, the separated flow gives rise to a drag force on the cylinder, with the drag force per unit length given by

$$F'_{\mathrm{d}} = \frac{1}{2}\rho U^2 D C_{\mathrm{d}}$$

where D is the cylinder diameter, U is the steady velocity of the flow, C_{d} is an empirical drag coefficient that depends only on the Reynolds number $\mathrm{Re} = UD/\nu$, and ν is the kinematic viscosity of the fluid.

For the second of these flows (Figure 8.8b), corresponding to an accelerating flow of an ideal fluid past a circular cylinder, there is no flow separation and so potential theory may be used. In this case, an inertia force proportional to the fluid acceleration is exerted on the cylinder, with the inertia force per unit length given by

$$F'_{\mathrm{i}} = \rho \frac{\pi D^2}{4} C_{\mathrm{m}} \dot{u}$$

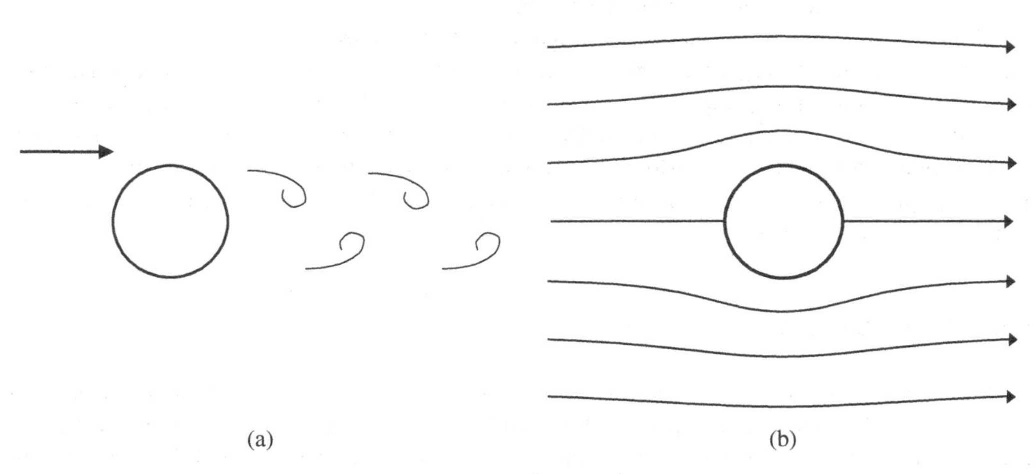

Figure 8.8 Reference flows used as a basis for development of the Morison equation. (a) Steady flow of a real fluid, (b) unsteady flow of an ideal fluid.

where $\dot{u}$ is the fluid acceleration and C_m is an inertia coefficient, with $C_m = 2$ for a circular cylinder. It should be noted that the formulation for the inertia force can be generalized to any cross-sectional shape and can be extended to three-dimensional bodies, with the cross-sectional area then replaced by volume. In such cases, the value of C_m is dependent on the shape of the two-dimensional or three-dimensional structure and the direction of acceleration. This flow is directly related to that resulting from an accelerating body in an otherwise stationary fluid, leading to the concept of "added mass," which is considered in Section 8.6.3.

The case under consideration corresponds to an accelerating flow of a real fluid past a cylinder section. This is treated by assuming a superposition of the two reference cases, but with a time-varying velocity u used in lieu of the steady velocity U and with C_d and C_m taking on empirical values. Hence, the Morison equation is developed as

$$F' = F'_d + F'_i = \frac{1}{2}\rho D C_d u|u| + \rho\frac{\pi D^2}{4}C_m\dot{u}$$

Note that $u|u|$ is used in place of u^2 in order to maintain the magnitude of the term, while assuring that the direction of the force changes with the direction of u (i.e. when u is negative, then so too is the corresponding drag force). Specific values of C_d and C_m need to be selected for any given situation. Typically, C_d ranges between about 0.6 and 1.2 and C_m ranges between about 1.4 and 2.0.

8.4.2 Morison Equation for a Sinusoidal Flow

When the Morison equation is applied in conjunction with linear wave theory, the incident flow past the section is sinusoidal, so that the case of a sinusoidal flow past a cylinder section is of particular interest. Consider a sinusoidal flow to be given by

$$u = u_o \cos(\omega t); \quad \dot{u} = -\omega u_o \sin(\omega t)$$

where u_o is the velocity amplitude and ω $(=2\pi/T)$ is the angular frequency. If these expressions for velocity and acceleration are substituted into the Morison equation, then the time-varying force can be expressed in the form:

$$X = A\cos(\omega t)\,|\cos(\omega t)| - B\sin(\omega t)$$

where X is the time-varying combined force, A is the maximum drag force, and B is the maximum inertia force. Thus,

$$X = F'$$

$$A = \frac{1}{2}\rho D C_d u_o^2$$

$$B = \rho\frac{\pi D^2}{4}C_m \omega u_o$$

Time variation of force. In order to illustrate the force variation with time, Figure 8.9 shows the time variation of the force components and combined force for an assumed sinusoidal variation in flow velocity.

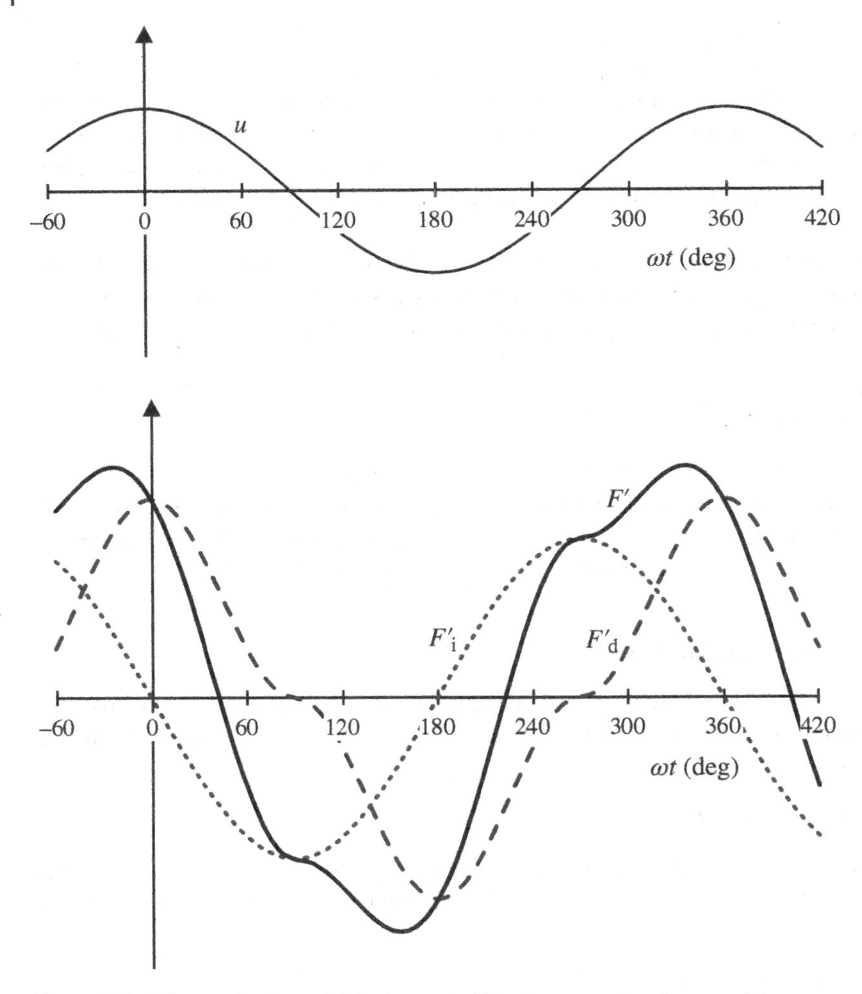

Figure 8.9 Time variations of velocity and of Morison equation force and force components.

The upper figure shows the time variation of the velocity u, while the lower figure shows the resulting time variation of the drag force (dashed line), inertia force (dotted line), and combined force (solid line). Thus, the drag force is in phase with the velocity and so is zero at times $t = T/4, 3T/4, \ldots$, and the inertia force is in phase with the acceleration and so is zero at times $t = 0, T/2, T, \ldots$.

Maximum force. For this case, the maximum force X_m is given as

$$X_m = \begin{cases} A + \dfrac{B^2}{4A} & \text{for } 2A > B \\ B & \text{for } 2A \leq B \end{cases}$$

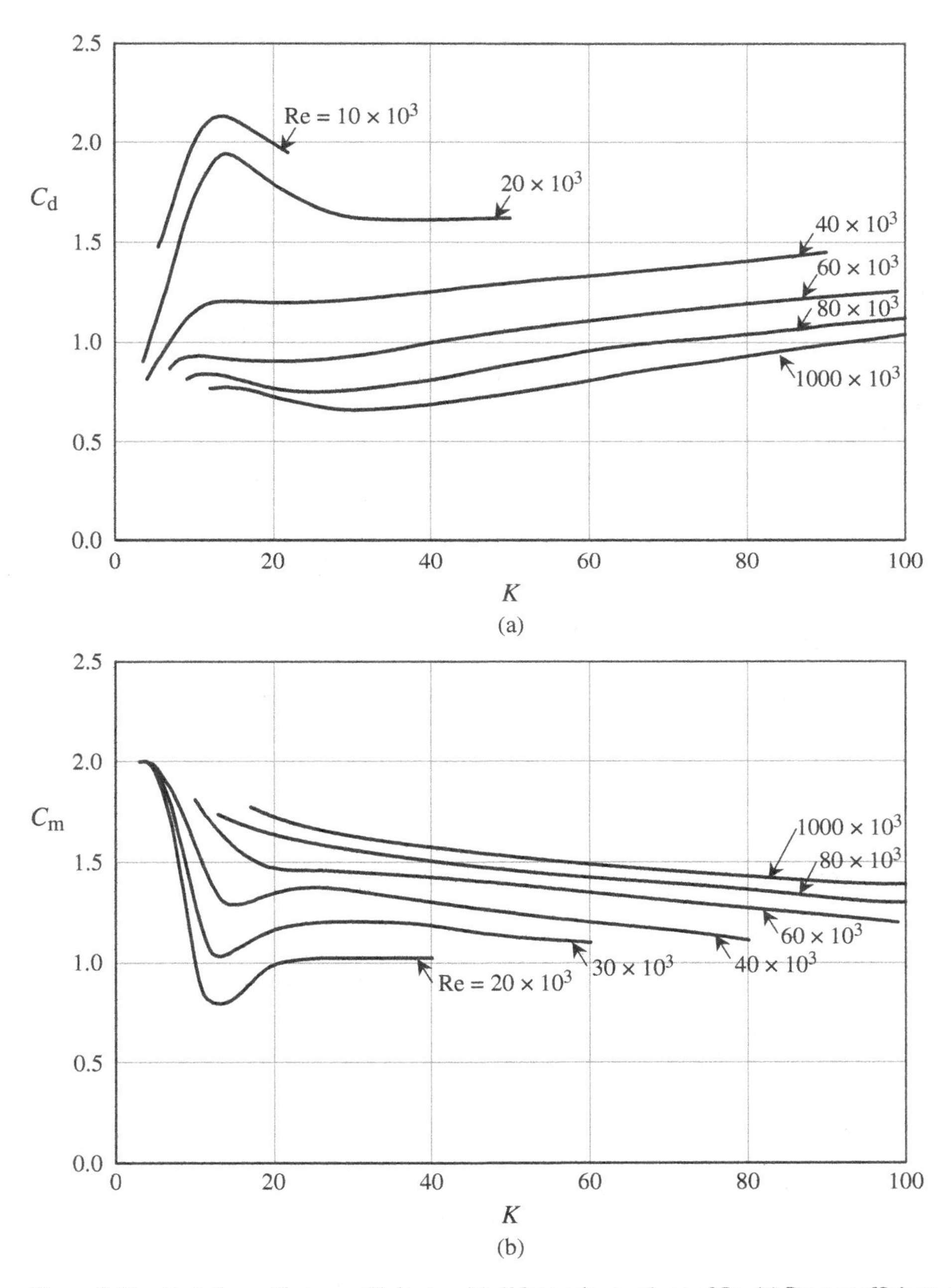

Figure 8.11 Variation of force coefficients with K for various values of Re. (a) Drag coefficient, (b) inertia coefficient.

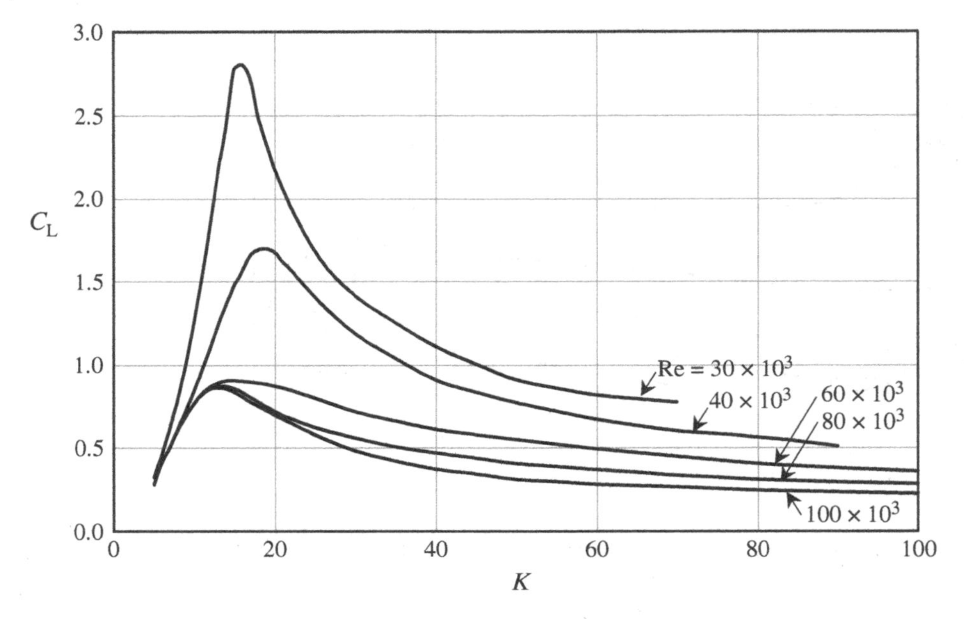

Figure 8.12 Variation of lift coefficient C_L with K for various values of Re.

- For $K < 5$: symmetrical flow, and therefore there is no transverse force
- For $K \simeq 5$–16: strong vortex shedding with $f_L/f_w = 2$, where f_w is the wave frequency; C_L exhibits a maximum with $C_L \simeq 1$–3 (depending on Re) at about $K = 12$
- For higher K: C_L reduces, with $f_L/f_w \simeq 0.2\,\mathrm{K}$

8.4.6 Extensions to the Morison Equation

A range of extensions to the Morison equation are possible. Some of these are now outlined.

Linearization. In applications of the Morison equation that involve random waves and/or a dynamic analysis of a structure, it is useful to develop a linearization of the Morison equation so that the total force is considered to be sinusoidal at the flow frequency. Such an approximation may be developed by utilizing the following expansion:

$$\cos\theta|\cos\theta| = \frac{8}{3\pi}\cos\theta + \frac{8}{15\pi}\cos 3\theta - \frac{8}{105\pi}\cos 5\theta + \ldots$$

In a linearization, only the first term on the right-hand side is retained, and the higher frequency terms are discarded. That is, the Morison equation is then approximated as

$$X = \frac{8}{3\pi}A\cos(\omega t) + B\sin(\omega t)$$

Nonlinear waves. The Morison equation may be applied in conjunction with a flow associated with nonlinear waves rather than a sinusoidal flow. In such a case, values of u and $\dot{u}$ are obtained from a nonlinear wave theory and applied at successive instants throughout a wave cycle. For a surface-piercing structure, integrations of the sectional force need to be carried out to the instantaneous surface elevation rather than to the SWL.

Waves with a current. The Morison equation may be applied in conjunction with a flow associated with waves combined with a current. That is, suitable values of u and $\dot{u}$ for the combined wave-current flow need to be determined and applied to the Morison equation.

Oblique structures. For a slender structural element whose axis is neither vertical nor horizontal, the flow velocity and acceleration to be applied to the Morison equation are those in a direction normal to the structure axis.

Complex structures. Numerical models for more complex structures comprising a number of slender elements may be developed, based on an extension of the case of an oblique element. This may be undertaken using a time-stepping method, with the loads obtained at each time step. At each instant, sectional forces on each element are obtained by the Morison equation in terms of u and $\dot{u}$ in a direction normal to the structure axis, with u and $\dot{u}$ obtained by linear or nonlinear wave theory, and the sectional forces are then summed numerically. Again, for a surface-piercing structure, integrations of the sectional force need to be carried out to the surface elevation rather than to the SWL.

Roughness elements. The roughness of the structure surface influences the loads acting on the structure. First, the effective diameter may be increased, and, second, the roughness elements affect the flow around the structure and so influence C_d and C_m. The roughness elements may be "hard" (e.g. barnacles or other hard growths on the structure surface) or "soft" (e.g. seaweed). For hard roughness elements of characteristic size k, information on the influence of the roughness parameter k/D on C_d and C_m is available.

Neighboring members. When cylindrical elements are adjacent to each other such that the wake from one element influences the flow around the other, the force on one element is influenced by the other, referred to as a shielding effect. This may occur when $\ell < d_\mathrm{o}$, where ℓ is the distance between the neighboring elements and d_o is the orbital diameter of the wave flow.

Moving structures. The Morison equation has been developed with respect to velocity u and acceleration u̇ of the flow relative to a structure that is fixed. When the structure itself is moving in the fluid with its own velocity $\dot{x}$ and acceleration $\ddot{x}$, then the Morison equation needs to be modified so as to account for the structure's motion. In such a case, the Morison equation for the force on a cylinder section is extended to

$$F' = F_\mathrm{d}' + F_\mathrm{i}' = \rho\frac{\pi D^2}{4}C_\mathrm{m}\dot{u} - \rho\frac{\pi D^2}{4}(C_\mathrm{m} - 1)\,\ddot{x} + \frac{1}{2}\rho DC_\mathrm{d}(u - \dot{x})|u - \dot{x}|$$

Numerical solutions of this nonlinear equation are available. In some instances, the structure motion itself arises from its equation of motion, e.g. as represented by a single-degree-of-freedom system, such that the structure motion is then influenced by the incident flow.

Fatigue analysis. There are examples of piles failing by fatigue when acted upon by waves and/or currents in a series of storms. To examine such situations, a fatigue analysis may be undertaken. In undertaking this, the Morison equation (typically the linearized version indicated above) is used in conjunction with a long-term distribution of individual wave heights, periods, and directions in order to obtain the number of cycles of different stress levels at a joint or structure section over a specified duration or the design life. The incorporation of a suitable description of the wave climate into a fatigue analysis is outlined in Section 4.7.4.

←■

8.5 Large Structures

8.5.1 Introduction

The case of a large structure for which a characteristic dimension D is no longer small relative to the wave length L is now considered. Because of its large size, the structure modifies the incident wave train by causing the waves to diffract around it, and this diffraction needs to be accounted for in determining the loads on the structure. However, in contrast to slender-member structures, the water particle orbits are now a small fraction of D, so that flow separation effects are absent or minor. This implies that linear diffraction theory, as introduced in Section 3.3 for the case of breakwaters, may now be used to develop the resulting flow. Wave loads for large structures have been described in detail by Sarpkaya and Isaacson (1981).

The velocity potential ϕ describing the flow is considered to represent the superposition of two flows: one corresponds to the incident wave field as if the structure was absent and the other is an additional flow that coexists with the incident wave field. Thus, ϕ is expressed as

$$\phi = \phi_{\mathrm{w}} + \phi_{\mathrm{s}}$$

The incident potential ϕ_{w} is known. The scattered potential ϕ_{s} is obtained as a solution to a boundary value problem that needs to account for a boundary condition on the structure surface, relating ϕ_{s} to ϕ_{w}, as well as a radiation condition in the far field. Closed-form and/or numerical solutions are available for most structure shapes.

8.5.2 Vertical Circular Cylinder

A notable reference case for which a closed-form solution is available corresponds to a vertical circular cylinder extending from the seabed to above the water surface. Figure 8.13 provides a definition sketch of this case.

The solution for this case is given in Table 8.2, which provides expressions for the velocity potential, the wave runup, and the loads on the cylinder.

In the table, complex expressions for the velocity potential and runup are given. (A complex representation of flow variables is indicated in Section 2.6.4.) Also, $i = \sqrt{-1}$, $\beta_{\mathrm{o}} = 1$, $\beta_{\mathrm{m}} = 2i^m$ for $m \geq 1$, C_{m} is an effective inertia coefficient, $H_{\mathrm{m}}^{(1)}(z) = J_{\mathrm{m}}(z) - iY_{\mathrm{m}}(z)$, J and Y are Bessel functions of the first and second kinds, respectively, each of order m and argument z, and a prime denotes a derivative with respect to the argument. Note that $J_1'(ka) = J_{\mathrm{o}}(ka) - J_1(ka)/ka$, and similarly for $Y_1'(ka)$. The effective inertia coefficient C_{m} and the phase angle δ are needed to determine the loads and their phases. Figure 8.14 provides these as a function of ka.

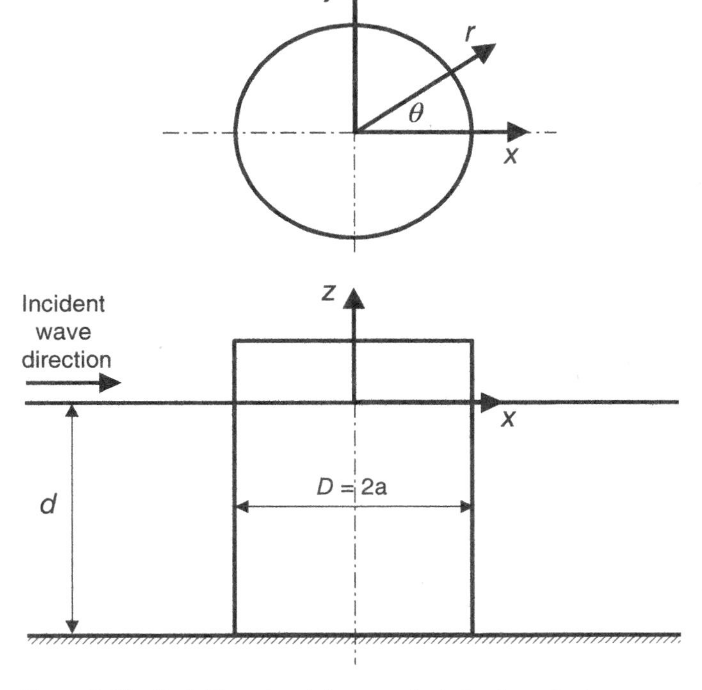

Figure 8.13 Definition sketch for a vertical circular cylinder.

Table 8.2 Closed-form solution for a large vertical circular cylinder.

Variable	Equation
Velocity potential	$\dfrac{\phi}{gH/\omega} = -\dfrac{1}{2}\dfrac{\cosh(ks)}{\cosh(kd)}\left\{\sum_{m=0}^{\infty} i\beta_m\left[J_m(kr) - \dfrac{J'_m(ka)}{H_m^{(1)\prime}(ka)}H_m^{(1)}(kr)\right]\cos(m\theta)\right\}\exp(-i\omega t)$
Runup	$\left(\dfrac{R}{H}\right) = \dfrac{1}{\pi ka}\left\lvert\sum_{m=0}^{\infty}\dfrac{i\beta_m\cos(m\theta)}{H_m^{(1)\prime}(ka)}\right\rvert$
Sectional force	$\dfrac{F'}{\rho gHD^2k} = \dfrac{\pi}{8}\dfrac{\cosh(ks)}{\cosh(kd)}C_m\cos(\omega t-\delta)$
Total force	$\dfrac{F}{\rho gHD^2} = \dfrac{\pi}{8}\tanh(kd)\,C_m\cos(\omega t-\delta)$
Overturning moment	$\dfrac{M}{\rho gHD^2d} = \dfrac{\pi}{8}\left[\dfrac{kd\sinh(kd)+1-\cosh(kd)}{kd\cosh(kd)}\right]C_m\cos(\omega t-\delta)$
where	$C_m = \dfrac{4}{\pi(ka)^2\sqrt{J_1'^2(ka)+Y_1'^2(ka)}}$
	$\delta = -\arctan\left[\dfrac{Y_1'(ka)}{J_1'(ka)}\right]$

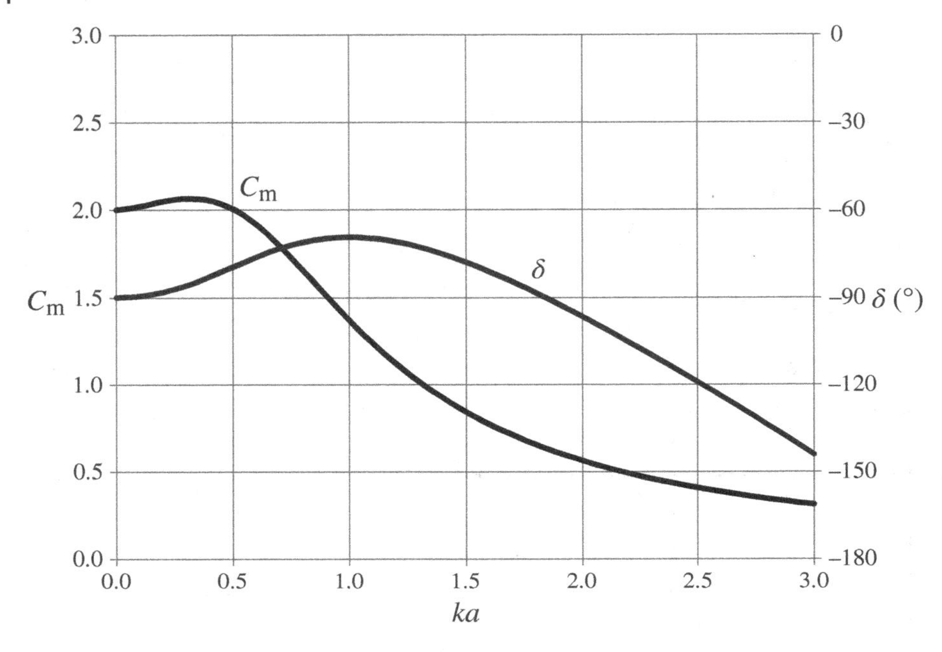

Figure 8.14 Variation of effective inertia coefficient C_m and phase angle δ with *ka*.

Reference Solution A9 in Appendix A provides a spreadsheet solution for the wave loads and runup for a large cylinder.

Example 8.4 ***Forces on Large Structures***
A surface-piercing, vertical, circular caisson of diameter $D = 10$ m in water of depth $d = 10$ m is subjected to an incident wave train with $H = 2$ m and $T = 5$ s. What is the maximum horizontal force on the caisson?

Solution

Specified parameters

$g = 9.80665\ \text{m/s}^2$
$\rho = 1025\ \text{kg/m}^3$
$H = 2.0\ \text{m}$
$T = 5.0\ \text{s}$
$d = 10.0\ \text{m}$
$D = 10.0\ \text{m}$

Solve for *kd*, *ka*

$a = D/2 = 5.0\ \text{m}$
$d/gT^2 = 0.041$

$kd = 1.72$
$ka = kd\,(a/d) = 0.86$

Effective inertia coefficient

C_m read off from Figure 8.14:
$C_m = 1.6$

Alternatively, C_m may be obtained from the formula in Table 8.2. Thus, Reference Solution A9 provides J_0, J_1, Y_0, and Y_1 values for $ka = 0.86$. Hence,

$$C_m = \frac{4}{\pi(ka)^2\sqrt{J_1'^2(ka)+Y_1'^2(ka)}} = 1.57$$

Force coefficient and force

$$C_F = \frac{\pi}{8}C_m \tanh(kd) = 0.58$$
$$F = \rho g H D^2 C_F = 1.2\text{ MN}$$

8.5.3 Other Configurations

More generally, numerical models based on alternative methods are available to obtain the wave field and the loads on a structure of arbitrary shape. These typically involve a discretization of the structure surface, the fluid domain near the structure, or the structure surface along with a portion of the fluid domain surface near the structure. In one such approach, the scattered potential ϕ_s is expressed in terms of a complex Green's function that is approximated as a series with N unknown coefficients that satisfies all the governing equations except for the structure surface boundary condition. The latter relates the scattered potential ϕ_s to the incident wave potential ϕ_w and is given as

$$\frac{\partial \phi_s}{\partial n} = -\frac{\partial \phi_w}{\partial n} \quad \text{at the structure surface}$$

where n is distance normal to the structure surface. The structure surface is discretized into N small elements and the boundary condition is applied at the N centers of the elements in order to obtain the N unknown coefficients. Once ϕ_s is known, it may be used along with the known ϕ_w to obtain the pressure distribution around the structure. Suitable integrations of this pressure distribution then result in the loads on the structure.

The above case of a structure of arbitrary shape in three dimensions may be simplified to alternate two-dimensional problems that involve a discretization of a line or profile rather than a surface. Thus, Figure 8.15 illustrates three kinds of two-dimensional problem that may be treated in this way. Figure 8.15a illustrates a horizontal cylinder of arbitrary section corresponding to a two-dimensional problem in the vertical plane, e.g. applicable to a section of a floating breakwater. Figure 8.15b illustrates a vertical cylinder of arbitrary section, corresponding to a two-dimensional problem in the horizontal plane, e.g. applicable to a rectangular-section vertical caisson. Figure 8.15c represents an axisymmetric structure, for which a Fourier-series representation of ϕ_s is used. In all three cases, only the structure profile needs to be discretized in order to solve the problem.

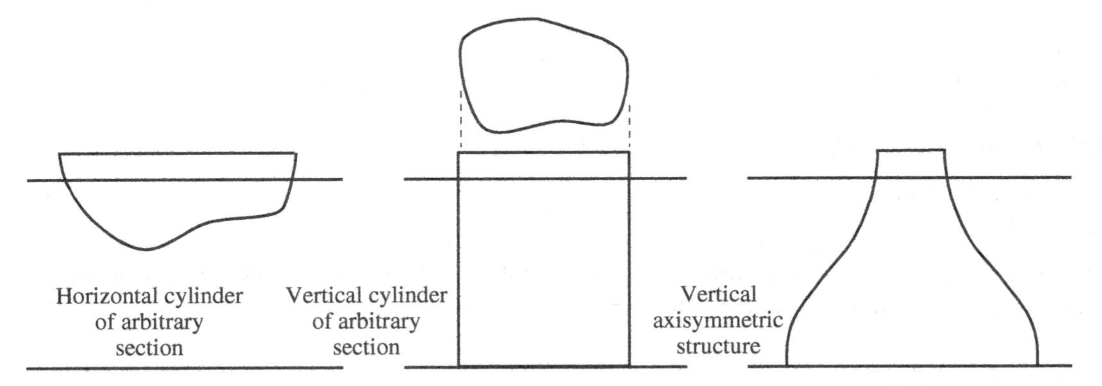

Figure 8.15 Illustration of structure configurations for which diffraction solutions are available. (a) Horizontal cylinder of arbitrary section, (b) vertical cylinder of arbitrary section, (c) vertical axisymmetric structure.

A range of other solutions are available for more specific configurations, such as a circular dock, semi-submerged and fully submerged horizontal circular cylinders, and a thin vertical barrier with a gap at the seabed.

8.6 Floating Structures

8.6.1 Introduction

It is possible to extend the concepts introduced in Section 8.5 with respect to a large structure so as to treat wave interactions with floating structures and the resulting structure motions. These may include floating breakwaters, bridges, docks, and – more relevant to naval architecture – marine vessels.

The general approach is to define the component motions of a floating structure, treat the boundary value problem for the flow potential so as to determine the hydrodynamic loads on the structure, and then solve the equations of motion taking account of the dynamic properties of the structure, its moorings or restraints if any, and the hydrostatic and hydrodynamic loads on the structure. The latter are associated in part with incident wave loads on the structure as if it was fixed and in part associated with fluid resistance forces associated with the structure motions.

Beyond an assessment of wave interactions with floating structures, it is also important to consider the design of a mooring system needed to restrain such a structure, as summarized in Section 8.8.1 in the context of floating breakwaters and bridges. In addition, the steady forward motion of a ship or marine vessel is a primary consideration in naval architecture, but is not treated here.

8.6.2 Recap of a Single-Degree-of-Freedom System

As relevant background to a determination of the motions of a floating structure, the behavior of a single-degree-of-freedom system is initially summarized. Figure 8.16 provides a definition sketch of a single-degree-of-freedom (sdof) system, also referred to as a spring-mass-dashpot system.

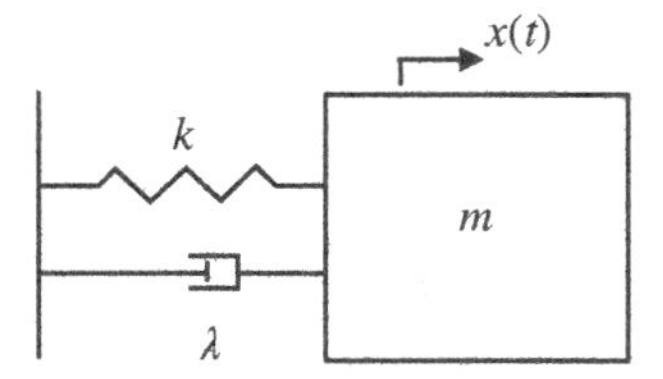

Figure 8.16 Definition sketch of a spring-mass-dashpot system.

Generally, the response x to an applied force F is to be determined. The system contains the following elements, with the force–response relationship for each of these indicated:

- mass, m: $F = m\ddot{x}$ (m has units of kg)
- damping, λ: $F = \lambda\dot{x}$ (λ has units of Ns/m)
- stiffness, k: $F = kx$ (k has units of N/m)

The equation of motion for the sdof system is

$$m\ddot{x} + \lambda\dot{x} + kx = F$$

The time variation of the excitation F may be sinusoidal, periodic, or take on some other specified form, or it may be random. However, only a sinusoidal excitation is considered here. The applied time-varying force F is given as

$$F = F_o \cos(\omega t)$$

where F_o is the force amplitude, t is time, and ω is the angular frequency of the excitation. ω is in units of rad/s, the corresponding circular frequency is given by $f = \omega/2\pi$ and is in Hz, and the corresponding period $T = 1/f = 2\pi/\omega$. The steady-state solution is given by

$$x = X \cos(\omega t - \delta)$$

where X is the amplitude of the displacement and δ is the phase angle between the force and the displacement. X and δ may be obtained from

$$\frac{X}{F_o/k} = \frac{1}{\sqrt{\left[1 - (\omega/\omega_n)^2\right]^2 + (2\zeta\omega/\omega_n)^2}}$$

$$\delta = \tan^{-1}\left[\frac{2\zeta\omega/\omega_n}{1 - (\omega/\omega_n)^2}\right]$$

In the above, the natural frequency ω_n is defined as $\omega_n = \sqrt{k/m}$

And the damping ratio ζ is defined as

$$\zeta = \frac{\lambda}{2\sqrt{mk}} = \frac{\lambda}{2m\omega_n}$$

Also, F_o/k represents the maximum displacement if the applied force is static and there are no dynamic effects. Therefore, the above ratio $X/(F_o/k)$ represents the relative increase in displacement associated with dynamic effects and is termed the dynamic amplification factor (DAF).

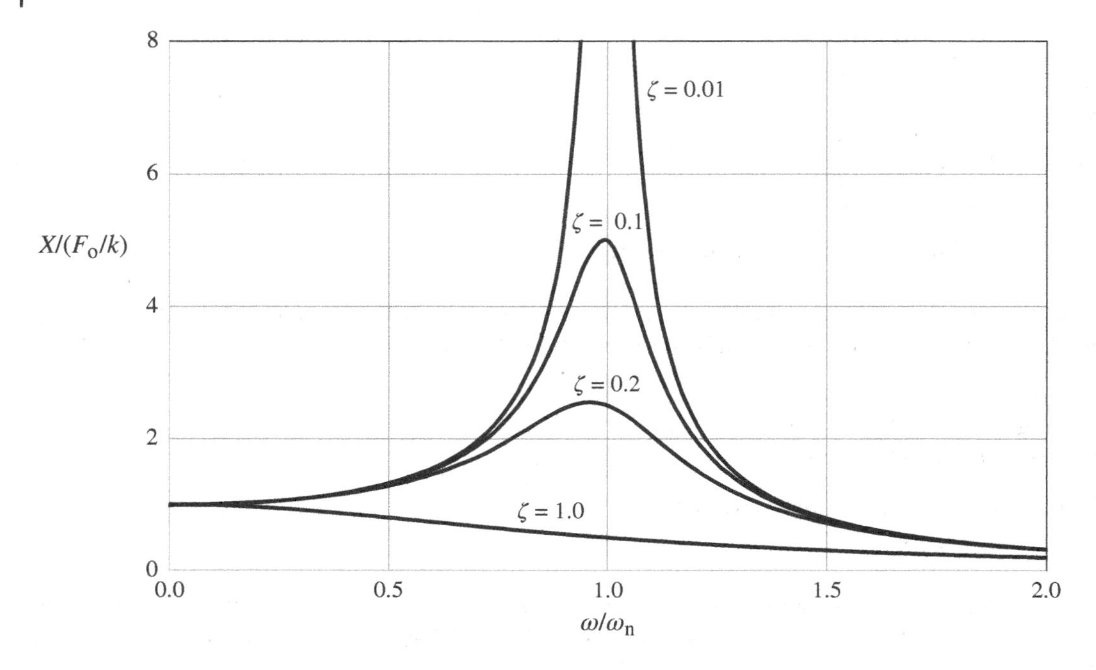

Figure 8.17 Response curves for a single-degree-of-freedom system.

Figure 8.17 shows the magnitude of the DAF as a function of the relative excitation frequency ω/ω_n for different damping ratios. For a lightly damped system, large resonant oscillations occur when the excitation frequency is close to the natural frequency. In the case of an undamped system ($\zeta = 0$), the response becomes infinite when $\omega/\omega_n = 1$.

8.6.3 Added Mass

Prior to considering the behavior of floating structures in waves, it is appropriate to outline the concept of added mass. A body accelerating in an infinite fluid causes some of the surrounding fluid to accelerate with it. This gives rise to a fluid force acting on the body in proportion to the body's acceleration. This force may be expressed as a fluid mass times the body's acceleration. This fluid mass is referred to as the added mass and may be considered to be added to the mass of the body itself in a force balance equation. The concept of added mass applies equally to a two-dimensional flow with a body accelerating normally to its axis and to a three-dimensional flow.

For any two- or three-dimensional body, the added mass is proportional to both the mass of the fluid displaced by the body and the body's acceleration. m'_a is used to denote the sectional added mass, i.e. the added mass per unit length for a two-dimensional flow, while m_a is used to denote the added mass for a three-dimensional flow. Thus, the added mass may be expressed as $m'_a = C_a \rho A \dot{u}$ for a two-dimensional flow and as $m_a = C_a \rho V \dot{u}$ for a three-dimensional flow, where A is the cross-sectional area of the body (in two dimensions), V is the volume of the body (in three dimensions), and C_a is an added mass coefficient that depends on the body configuration and the direction of acceleration.

Added mass coefficients for two- and three-dimensional flows may be obtained by analytical or numerical methods used to treat the corresponding potential flows and are available in the literature for various body shapes and acceleration directions (e.g. Sarpkaya and Isaacson 1981). Specific values of the added mass coefficient include the following. In two dimensions, $C_a = 1$ for a circular section and $C_a = 1.19$ for a square section accelerating in a direction normal to a side. In three dimensions, $C_a = 0.5$ for a sphere and $C_a = 0.7$ for a cube accelerating in a direction normal to a face.

When extending this concept to a floating body with a free surface, the accelerating body also creates a wave system on the water surface and so the resulting added mas is then frequency dependent.

8.6.4 Hydrodynamic Analysis

The general approach to addressing the wave loads on, and the response of, a large floating structure initially requires a distinction between two related problems:

- one is the *diffraction problem*, which corresponds to incident wave interactions with a fixed structure, in the manner considered already in Section 8.5; and
- the other is the *radiation problem*, which corresponds to the structure oscillating in otherwise still water, resulting in fluid loads that are proportional to the structure's motions.

The floating structure case is treated by a combination of these two problems, with the forces from the diffraction problem, termed exciting forces, causing the structure motions, while the structure motions generate additional forces on the structure that resist the motions. With respect to the latter, the load components in phase with the structure's velocity are expressed in terms of damping coefficients, while the components in phase with the structure's acceleration are expressed in terms of added masses (an introduction to the concept of added mass is given in Section 8.6.3).

Equations of motion. The case of a structure undergoing heave motions only (in the vertical direction) is initially considered. The equation of motion may be written as

$$(m + \mu)\ddot{x} + \lambda\dot{u} + kx = F_e$$

where m is the mass of the structure, μ is the added mass (i.e. the corresponding force on the structure is $-\mu\ddot{x}$), λ is the damping coefficient (i.e. the corresponding force on the structure is $-\lambda\dot{x}$), k is a stiffness (i.e. the corresponding force on the structure is $-kx$), and F_e is the exciting force. The above equation is analogous to the sdof equation of motion presented earlier. F_e is obtained as a solution to the boundary value problem in the manner indicated in Section 8.5, while μ and λ can be obtained as solutions to a closely related boundary value problem, the radiation problem, in which the structure surface boundary condition is related to the structure motions rather than to the incident wave field.

Such an approach may be extended to a rigid structure undergoing component motions with several degrees of freedom. In general, a rigid body may oscillate with six degrees of freedom as indicated in Figure 8.18. Thus, using well-known nautical terms, the displacements in the x, y, and z directions are referred to as surge, heave, and sway, respectively, and the rotations about the x, y, and z axes are referred to as roll, pitch, and yaw, respectively.

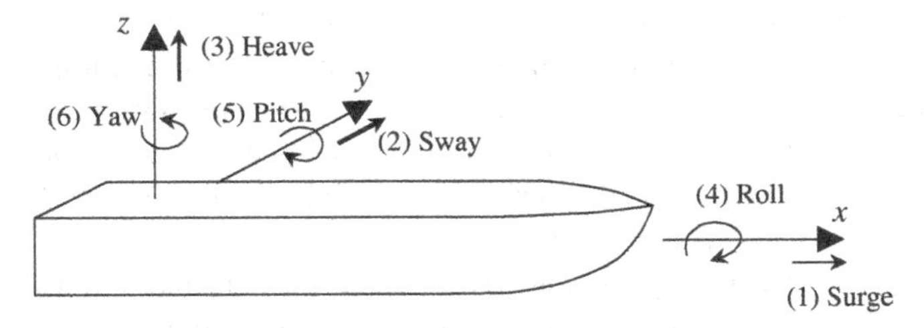

Figure 8.18 Definition sketch of structure motions.

Taking these into account, the above approach may be extended to six coupled equations of motion in lieu of a single equation of motion:

$$\sum_{j=1}^{6}(m_{ij} + \mu_{ij})\ddot{x}_j + \lambda_{ij}\dot{x}_j + k_{ij}x_j = F_i \quad \text{for } i = 1, 2, \ldots, 6$$

where x_j with $j = 1, \ldots, 6$, represents the six component motions to be determined, F_i are the six exciting force components, and m_{ij}, μ_{ij}, λ_{ij}, and k_{ij} are the cross components of mass, added mass, damping coefficient, and stiffness, respectively. F_i, μ_{ij}, and λ_{ij} are obtained as solutions to the diffraction/radiation problems, leading to a solution for x_j. In presenting such results, motion amplitudes are commonly expressed as *response amplitude operators* that correspond to the motion amplitudes per unit wave amplitude expressed in dimensionless form. These are provided as functions of frequency.

Mass and stiffness matrices. The mass matrix terms m_{ij} and the hydrostatic stiffness matrix terms k_{ij} in the equations of motion are generally known for a given body configuration. For the two-dimensional case of a body section that is symmetrical about the z axis, the mass and stiffness matrices are given, respectively, as follows:

$$\begin{array}{cccc} & (sway & heave & roll) \\ m_{ij} = & \begin{bmatrix} m' & 0 & -m'z_G \\ 0 & m' & 0 \\ -m'z_G & 0 & I'_o \end{bmatrix} \end{array}$$

$$\begin{array}{cccc} & (sway & heave & roll) \\ k_{ij} = & \begin{bmatrix} 0 & 0 & 0 \\ 0 & \rho g B & 0 \\ 0 & 0 & (\rho g S_{22} - m'g\,\mathrm{BG}) \end{bmatrix} \end{array}$$

where m' is the mass per unit length (in units of kg/m), B is the beam, $\mathrm{BG} = z_G - z_B$, z_G and z_B are the z coordinates of the centers of gravity and buoyancy, respectively, I'_o is the moment of inertia about the origin per unit length, which is given as $I'_o = m'\left(r_y^2 + z_G^2\right)$, and r_y is the radius of gyration in the x–z plane. Also, $S_{22} = B^3/12$ is the waterplane moment of inertia per unit length.

Haskind relations. These refer to certain relationships between the scattered and radiated potentials that can be used to develop relationships between the exciting forces and damping coefficients. They are

Example 8.5 ***Floating Breakwater Performance***
A rectangular-section floating breakwater with beam $B = 6$ m and draft $h = 1.2$ m in water of depth $d = 5$ m is constrained by piles so as to undergo heave motions only. Estimate the natural period in heave. Assuming the breakwater does not move, estimate the transmission coefficient for a wave period $T = 2.5$ s.

Solution

Specified parameters

$g = 9.80665\ \text{m/s}^2$
$B = 6.0$ m
$h = 1.2$ m
$T = 2.5$ s
$d = 5.0$ m

Natural period in heave

Parameters needed to apply the formula for T_h:

$B/h = 5.0$
$C_h = 0.552 + 0.152(B/h) = 1.3$

From formula for T_h:

$T_h = 3.3$ s

Transmission coefficient

Additional parameters needed to apply the Macagno formula for K_t:

$d/gT^2 = 0.0816$
$kd = 3.23$
$L = 2\pi d/kd = 9.7$ m
$B/L = 0.62$
$h/d = 0.24$

From the Macagno formula for K_t:

$K_t = 0.23$

■➔

8.8.2 Hydrodynamic Analysis

If a more complete hydrodynamic analysis is warranted, which is invariably the case for a floating bridge, this may be undertaken in the manner described in Section 8.6.4. Thus, for normally incident waves, the equations of motion for three degrees of freedom (sway, heave, and roll) are

$$\sum_{j=1}^{3}(\lambda_{ij} + \mu_{ij})\ddot{x}_j + \lambda_{ij}\dot{x}_j + k_{ij}x_j = F_i \quad \text{for} \quad i = 1, 2, 3$$

where the notation of Section 8.6.4 is used. The mass and stiffness elements, m_{ij} and k_{ij}, are determined from the structural/hydrostatic properties of the structure section and the exciting force components F_i, the added masses μ_{ij}, and the damping coefficients λ_{ij} are obtained from a solution to the diffraction/radiation problem in the manner described in Section 8.6.4.

Effects of obliquely incident waves are especially important for floating bridges. These cause differential stresses along the bridge length and therefore need to be examined in order to assess the structural integrity of the bridge, especially at connections between individual units. In a more advanced analysis, wave directional spreading is accounted for, so that the incident wave field is represented as a directional wave spectrum.

In a numerical simulation, the directional spectrum is represented as a double summation of component wave trains over frequencies and directions (see Section 4.3.3), the bridge structure is discretized into a number of segments, and a hydrodynamic analysis is used to obtain force coefficients and corresponding phases for each component wave train. Thus, the force on the ith bridge segment is obtained from a double summation over frequencies (index j) and directions (index k) in the following manner:

$$F_i(t) = \sum_j \sum_k a_{jk}\, C_{ijk}\, \cos(k_j y_i \sin\theta_k - \omega_j t - \varepsilon_{jk} - \phi_{ijk})$$

In this equation, y_i is the y coordinate measured along the bridge axis at the center of ith bridge segment, C_{ijk} and ϕ_{ijk} are force coefficients and phases, respectively, obtained from the hydrodynamic analysis, and a_{jk} and ε_{jk} are respectively the amplitudes and phases of the component waves used to develop the wave spectrum simulation. The force components on a section of the bridge are then obtained by integrating the force components over the section length.

Wave drift force. In addition to the first-order oscillatory forces acting on a structure, a steady second-order force, termed the wave drift force, is also present. For the two-dimensional case of normally incident regular waves with no energy dissipation, the wave drift force F_{D} is given by

$$F_{\mathrm{D}} = \frac{1}{8}\rho g H_{\mathrm{i}}^2 \left[1 + \frac{2kd}{\sinh(2kd)}\right] K_{\mathrm{r}}^2$$

where H_{i} is the incident wave height, K_{r} is the reflection coefficient, and kd is the depth parameter. Variations of this formula that take account of wave energy dissipation and obliquely incident waves are available. The wave drift force is especially important for random waves incident on a moored structure or vessel, since the drift force will then vary slowly with time and this low-frequency variation may excite a low-frequency resonance of a structure's mooring system.

8.8.3 Mooring System Analysis

The mooring system of a floating breakwater, floating bridge, moored ship, or other marine vessel is generally designed to limit mean offsets and drifting motions, while allowing for tidal elevation changes and for the structure's oscillatory motions. A mooring system design considers anchor type and weight, the line weight per unit length, the number of lines along with the length of each line, the horizontal extent and orientation of each line, and their attachments to the structure. A mooring analysis may be undertaken to varying degrees of sophistication, including a simpler static mooring analysis and a more complete dynamic mooring analysis for motions with six degrees of freedom. These analyses are considered in turn.

8.8.3.1 Static Mooring Analysis

In a simpler static analysis, the mooring system may be analyzed by considering the maximum offset of the structure. First, the mean horizontal offset is obtained by balancing the combined horizontal force of all the cables against the wave drift force, the current drag force, and the wind drag force. This mean horizontal offset is then taken to include additional horizontal and vertical offsets due to the oscillatory structure motions.

The mooring analysis then provides the cable profiles, cable tensions, anchor loads, and loads at the connection points. A more complete analysis involves the catenary equations for each cable, taking account of multiple cables, cables partially resting on the seabed, and possibly the elasticity of the cables and floats or weights attached to the cables. The catenary equations are nonlinear, but may be solved numerically for any specified conditions.

Analysis of a single inextensible cable. While a full numerical solution for a given mooring system is normally relied upon, it is useful to illustrate the fundamentals of this approach with respect to an analysis of a single inelastic cable. The cable is assumed to be uniform and inextensible (reasonable for steel chain), with no floats or weights attached, and it may be fully or partially suspended. Figure 8.22 provides a definition sketch for this case, indicating the various loads acting on the cable.

In the figure, q is the buoyant weight per unit length of cable, C is the total length of the cable, C_s is the length of the suspended portion of the cable, L is the horizontal distance between the anchor and the upper attachment point, and L_s is the horizontal span of the suspended portion of the cable. Also, h is the vertical distance between the upper attachment point (to the structure) and the lower attachment point (to the anchor) or the point of contact with the seabed, T is the cable tension with horizontal and vertical components H and V, respectively, subscript 1 is the upper attachment point, subscript 0 is the lower attachment point or the point when the cable just touches the seabed, and the arrows denote loads. H remains constant along the suspended length of the cable, while T and V vary along the cable.

Typically, H, C, L, and q are assumed to be known, and a computation is undertaken to provide T and V (variable along the cable), H (constant along the cable), and the cable profile. The relevant equations are

$$\frac{C}{L} = \sqrt{[q \sinh(q)]^2 + h^2}$$

$$\tan(\theta) = L\beta \, [h \coth(\theta) - C]$$

where $\beta = qL/2H$

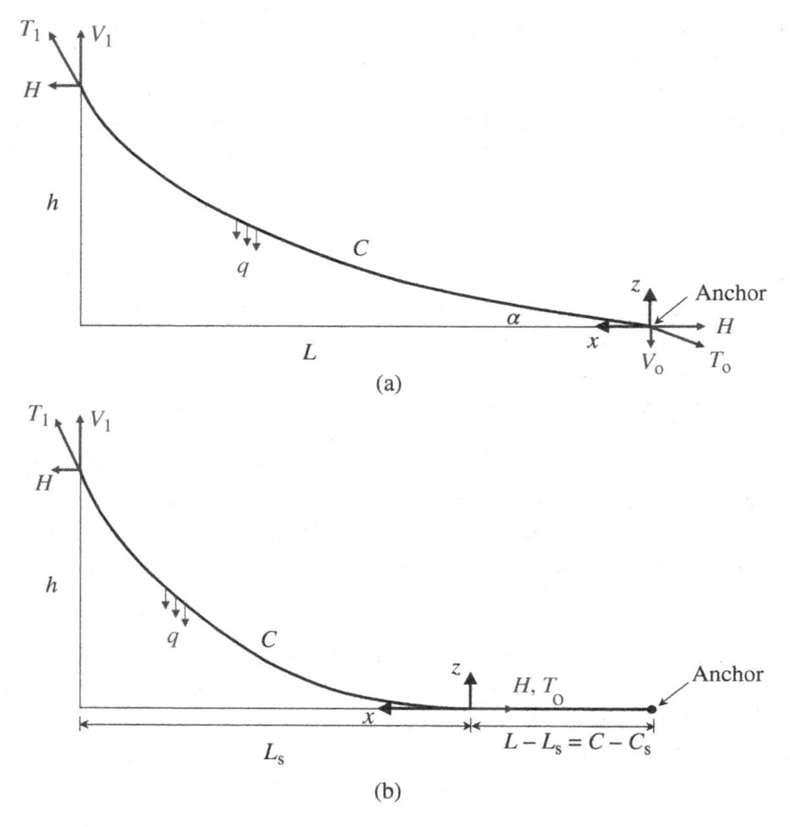

Figure 8.22 Definition sketch for loads on fully and partially suspended cables. (a) Fully suspended cable, (b) partially suspended cable.

If $\tan(\theta) < 0$, then the cable is only partially suspended. In such a case, $\theta = 0$ is assumed, but the calculation is repeated with C and L replaced by C_s and L_s, respectively. Also, the following is needed:

$$C_s = L_s + C - \mathrm{L}$$

The above equations may be solved numerically for any particular conditions, often based on a finite element method.

8.8.3.2 Dynamic Mooring Analysis

More commonly, a more complete dynamic analysis of a mooring system may be undertaken involving a sophisticated mooring analysis model. Typically, such a model would solve the time-domain equations of motion in six degrees of freedom, taking into account forces from a specified wave field, current field, and wind field. Such models allow for a wide range of mooring configurations and types of floating structures. The model results include the temporal variations of the structure motion in six degrees of freedom, along with the associated mooring line loads, and loads at the anchors and connection points.

←■

8.9 Other Loads

While the focus of this chapter has been on wave loads on coastal structures, a selection of other potential forms of loading is now summarized.

8.9.1 Foundation Loads and Stability

In caisson design, an assessment of a caisson's foundation and the caisson's associated stability is required. This is considered to lie within the field of geotechnical engineering. (Other geotechnical topics relevant to coastal structures relate to pile design and anchor design.) For caissons, the weight of the caisson and any superstructure, buoyancy forces, and environmental loads due to wind, waves, currents, and, possibly, earthquakes need to be accounted for. A stability assessment involves a range of considerations such as overturning, horizontal sliding, settlement, bearing capacity, and liquefaction. Each of these is summarized below:

- ***Overturning***. This is initiated when the overturning moment due to environmental loads exceeds that due to the submerged weight of the caisson. Since the wave load is oscillatory, this may lead to rocking behavior rather than full overturning.
- ***Sliding***. This refers to the horizontal movement that occurs because horizontal forces exceed the resistive capacity of the soil.
- ***Settlement***. This occurs when a caisson settles below the undisturbed seabed level. This may include elastic settlement (depending on the elastic properties of the soil) and settlement due to consolidation (in which soil particles pack together more tightly).
- ***Bearing capacity***. This refers to the capacity of the soil to support the loads applied to it through the structure. If the bearing capacity is exceeded because of excessive pressures, a shear failure occurs, leading to an unduly large settlement.
- ***Liquefaction***. Liquefaction may occur in saturated soil (which, of course, is present at the seabed), when external oscillatory loading can cause water pressures within the soil to increase to the extent that soil particles readily move in relation to each other, causing a significant reduction in soil strength. While earthquake-induced liquefaction is more commonly considered, wave-induced liquefaction is also possible.

8.9.2 Earthquake Loads

Earthquake loads may be estimated using well-established approaches for the seismic design of land-based structures. This is considered to lie within the field of earthquake engineering. The key distinction for a submerged structure is that, since the vibrating structure causes some of the surrounding fluid to accelerate with it, the fluid force acting on the structure corresponds to the structure acceleration multiplied by the added mass, so that the added mass can be considered to be added to the mass of the structure itself in a structural analysis.

Vertical wall. The reference case of a vertical seawall is initially considered. The flow is assumed to be two-dimensional in the (x, s) plane, where x is distance normal to the wall and s is distance measured upwards from the seabed. That is, a portion of a wall that is infinitely long is considered. Westergaard developed a solution for this case in 1931 in the context of a dam with a vertical face. The simplified Westergaard solution provides the vertical distribution of added mass per unit height, $\mu'(s)$, the added mass μ, and the effective elevation h above the seabed at which the added mass is effectively located. These are given respectively by

$$\mu'(s) = \frac{7}{8} \rho w \sqrt{ds}$$

$$\mu = \frac{7}{12} \rho w d^2$$

$$h = \frac{2}{5} d$$

where d is the water depth and w is the selected width of the wall. In the case of a two-sided wall with water on both sides, the water on both sides is made to accelerate, so that the added mass is twice that of a one-sided wall.

***Vertical circular cylinder*.** A second reference case corresponds to a vertical circular cylinder extending from the seabed to above the water surface. For this case, the high-frequency added mass μ for horizontal motions of the cylinder is given by the expression

$$\frac{\mu}{\rho a d^2} = -\frac{16}{\pi^2} \sum_{m=1}^{\infty} \frac{1}{(2m-1)^3} \frac{K_1(k_m a)}{K_1'(k_m a)}$$

where a is the cylinder radius, d is the water depth, $K_\nu(z)$ is the modified Bessel function of the second kind with order ν and argument z (available via a spreadsheet), $K_1'(z) = -K_0(z) - K_1(z)/z$, and $k_m a$ is given by

$$k_m a = (2m-1) \frac{\pi}{2} \frac{a}{d}$$

The solution indicates that $\mu/\rho a d^2$ increases with increasing a/d, reaching a limit of 1.7 for large values of a/d.

8.9.3 Vessel Impact, Ice Impact, and Debris Loads

Vessel impact. Vessel impacts on berths, caissons, and other coastal structures may be estimated through momentum considerations taking account of an assumed mass, added mass, velocity, and trajectory of a vessel.

Ice mass impact. Methods of determining ice loads on coastal structures fall within the discipline of ice engineering. Floating ice masses include ice floes, bergy bits (smaller ice masses rising 1–4 m above the water surface), or growlers (ice masses rising no more than 1 m above the water surface). Icebergs may be a consideration for offshore structures in deeper water but are not especially relevant for coastal structures because of grounding in shallow water. Loads due to a floating ice mass are dependent on ice properties, the structure and ice mass configurations, the ice mass trajectory, the drift velocity due to waves and currents, and wave-induced oscillatory velocities if present. A hydrodynamic analysis, similar to that for a floating offshore structure, may be used in combination with an ice impact model.

Ice floe impact. Distinct from impacts from floating ice masses, loads due to a moving ice floe or sheet of ice are dependent on ice properties, sheet velocity, and the structure configuration as affecting ice ride-up, ice crushing, and ice fracture mechanisms. For such cases, a structure with a sloping surface is usually advantageous in causing the sheet of ice to deflect up the structure surface and so fail more readily by bending rather than by crushing.

Debris loads. Coastal structures may be impacted by floating debris during flood or storm events, including logs and unmoored small boats – and even vehicles in the case of tsunami flooding. Design guides (e.g. FEMA 2009; FEMA 2011) describe the estimation of debris loads based on assumptions of debris mass, sometimes limited by water depth, and debris velocity. There is a distinction between smaller debris that can be moved by wave motions and so impact a coastal structure at relatively high velocities, and larger debris that are moved more slowly by local currents, but are not moved at high velocities by waves.

8.9.4 Wind Loads

Methods of determining wind loads fall within the field of wind engineering. Considerations include design wind speeds, usually accounting for the wind profile with elevation, the steady drag force based on assumed drag coefficients for different structural configurations, and, if necessary, the dynamic response of a structure arising from an oscillating lift force and possibly from the oscillatory component of a drag force.

Wind loads may be relevant in coastal engineering with respect to, for example, mooring system design of ships and floating docks and wind loads acting on ships, boats, and other marine vessels that result in increased loads on docks and other facilities.

8.10 Renewable Energy Infrastructure

8.10.1 Background and Criteria

While this chapter has focused on coastal structures, this section provides a supplement to such considerations in order to outline renewable energy infrastructure relating to ocean energy sources. It is recognized that the design of such infrastructure, which includes structures, facilities, and devices, encompasses many branches of engineering, including ocean engineering. (Extraction of nonrenewable forms of energy from the ocean, most notably oil and gas, is central to the discipline of ocean engineering, but is not considered in this text.)

There is an increasing reliance on renewable energy sources worldwide, based on society's recognition of climate change impacts and the need to minimize impacts on the environment. Thus, significant effort has been expended over several decades to examine, develop, and commercialize the extraction of renewable energy from the oceans. Renewable energy refers to energy from natural sources that is not depleted but is replenished as it is extracted, even over the long term. Along the shorelines and in the ocean, renewable energy sources may derive from the wind, waves, tides, currents, and seawater temperature differences. The key criteria for the development and commercialization of a successful technology include economic feasibility (although subsidies are typically provided until the technology is proven to be fully viable), structural stability and integrity with respect to extreme storms, constructability, operability including the need to minimize maintenance requirements, reliability with respect to minimizing disruptions, and the need to minimize impacts on the environment.

Approaches that seek to extract energy from these five sources are now outlined. Apart from wind energy extraction, which is widespread, approaches reliant on the other energy sources have encountered technological challenges and/or have not yet proven to be commercially viable.

8.10.2 Wind Energy

A wind farm refers to a group of wind turbines at the same location. The technology for onshore wind turbines is now well developed, with land-based wind farms now generating a significant amount of renewable energy worldwide.

Key benefits associated with the development of offshore wind farms include the following: they rely on established technology, based on the success of onshore wind farms; offshore winds represent an abundant renewable energy source, with stronger winds offshore than overland, leading to greater efficiencies; and the distribution of available energy through the year leads to efficiencies, with stronger winds in the winter when energy needs are higher. On the other hand, relative to land-based wind turbines, offshore wind turbines are more costly to install, access, and maintain, and they need to be sufficiently robust to withstand extreme storms. While turbines in shallower water may be bottom-founded, those in deeper water need to be mounted on floating platforms, with relatively expensive designs that take account of mooring systems and platform motions.

Offshore wind farms are now relied upon extensively worldwide.

8.10.3 Wave Energy

Ocean waves possess significant amounts of potential energy associated with a changing water surface and kinetic energy associated with fluid velocities. Many devices have been proposed or developed to extract energy from ocean waves.

Potential issues. Key issues that require consideration in the development of a wave energy device include the following. Bearing in mind the variability in wave climate throughout a year and the complexity of wave flows, with wave-by-wave irregularities, multiple frequencies, multiple directions, and potential wave breaking, there is a need to ensure that the device is reasonably efficient over a range of wave frequencies, heights, and directions. Although larger storms give rise to the greatest rates of energy capture, the structure needs to be able to withstand the most severe storms that occur. As with other forms of energy, there is a need to achieve economic feasibility and minimize maintenance requirements, in part by minimizing the complexity of mechanical systems (e.g. the number of moving parts).

Categories of devices. These devices may be categorized by their relative location: they may be located along the coastline, in the nearshore where they may be bottom-founded, or offshore where they are moored or tethered. They may also be categorized by the method of energy capture: this is usually achieved either by the relative motion of adjacent components used to drive a hydraulic piston or by the development of a low-head differential used to drive a generator.

Specific devices that have been proposed or developed include the following:

- a buoy designed such that the rising and falling water surface drives hydraulic fluid;
- a long surface-following device with sections that articulate with the movement of the waves, with the relative motion between adjacent sections used to create pressurized fluid;
- a piston pump attached to the seabed with a float tethered to the piston;
- a hinged mechanical flap connected to the seabed and extending to the water surface, with the flap movement driving a hydraulic piston; and
- a shore-based device in which waves run up a ramp and overtop into a low-head reservoir, with the water return driving a generator.

However, while many devices have been proposed or developed, with some connected to the power grid and deployed over multiple years, to date none have been found to be sufficiently reliable or economically feasible, so as to assure widespread, long-term application.

8.10.4 Tidal Energy

Tidal fluctuations possess significant kinetic energy associated with tidal currents, especially through narrow channels, and significant potential energy associated with a changing water surface elevation. Infrastructure takes the form of either current turbines or tidal barrages, which are dam-like structures that make use of a differential head between high and low tides.

Other undeveloped approaches include the use of tidal lagoons (a barrage at a manmade location), which is referred to as dynamic tidal power. This relies on a head differential across a barrier or effectively a dam extending from the shoreline over a long distance, so that a head differential across the dam arises as the tidal current moves parallel to the shoreline.

Current turbines are considered in Section 8.10.5, while tidal barrages are now considered. Key issues that require consideration in the development of a tidal barrage include economic feasibility, bearing in mind their high capital costs, the limited availability of sites with a sufficiently high tidal range, and negative environmental impacts with respect to marine life and habitat.

Only two large-scale tidal power plants with capacities of over 200 MW have been completed and are operational. These consist of a series of sluice gates and a series of generators, both built into the barrage. Through one part of a tidal cycle, the gates are opened to enable the flood tide to fill the reservoir, followed by the gates being closed and the turbines opened so as to enable the differential head to drive the turbines.

8.10.5 Current Turbines

Strong currents associated with a particular bathymetry or shoreline configuration, often associated with a tidal flow, may be used to drive a current turbine. The design of current turbines is distinct from the design of wind turbines on account of the greater fluid density and lower flow velocity. Key issues that require consideration include economic feasibility, potential negative environmental impacts with respect to marine life and habitat, and the technology needed to capture power from relatively low current speeds. The most common devices correspond to horizontal axis propellors that may be open or ducted, while vertical axis turbines have also been studied. Although a few commercial-scale grid-connected current turbines have been deployed, none have been found to be sufficiently reliable or economically feasible so as to assure widespread, long-term application.

8.10.6 Ocean Thermal Energy Conversion

An ocean thermal energy conversion (OTEC) system relies on temperature differences between cooler deep seawater and warmer near-surface seawater to run a heat engine. A key consideration is that the temperature differential is small and the thermal efficiency is low. When this issue is combined with the major capital investments that are needed, the economic feasibility of such a system remains a challenge. On the other hand, technological challenges associated with a large, moored floating offshore structure have largely been addressed in the context of oil and gas exploration and recovery.

While a few on-land, small-scale demonstration plants have been developed to date, the prospect of full-scale offshore OTEC plants that produce power at a commercial scale appears to be remote.

Problems

[Where relevant, you should rely on the values of physical constants provided in Appendix C.]

8.1 A regular wave train with height $H = 1.4$ m and period $T = 4$ s approaches a vertical seawall normally in water of uniform depth $d = 4$ m. The wall is to be constructed in two sections: one extending from the seabed to mid-depth $s = 2$ m, and the other above $s = 2$ m. Estimate the maximum force and overturning moment per unit width on the entire wall using the Miche-Rundgren/Sainflou method. ■ Estimate, also, the maximum force per unit width on the lower section of the wall using the Miche-Rundgren/Sainflou method.

8.2 A wave train with significant wave height $H_s = 0.8$ m and period $T = 4.0$ s obliquely approaches a vertical seawall in water of uniform depth $d = 4$ m. The wave crests make an angle of 25° with the wall, and the top of the wall is 5 m above the seabed. Using Reference Solution A8 in Appendix A, estimate the maximum force and maximum overturning moment per unit width on the wall on the basis of the Goda formulation. What would these values be if the effects of wave breaking are omitted?

8.3 A rubble-mound breakwater with a 1/1.5 slope is to be located at a site where the design wave height is $H_s = 1.5$ m. On the basis of the Hudson equation, estimate the median mass and nominal diameter of rock armor units required for no damage to the breakwater. Assume that the rock has a density $\rho_s = 2650$ kg/m and that the stability coefficient $K_D = 4.0$. Estimate the nominal diameter for the above conditions based instead on the van der Meer equations, assuming that the zero-crossing period $T_z = 5$ s, the permeability factor $P = 0.5$, and the design storm duration $\tau = 6$ hr.

8.4 A two-dimensional oscillatory flow past a submarine pipeline of diameter $D = 1$ m has a maximum velocity of 2.2 m/s and a period of 6 s. Assuming that $C_d = 1.0$ and $C_m = 2.0$, what is the maximum in-line force per unit length on the pipe? Estimate when this occurs, measured in seconds before the flow velocity is a maximum.

8.5 Estimate the maximum in-line force and overturning moment acting on a vertical pile of diameter $D = 0.5$ m in water of depth $d = 6$ m when it is subjected to waves of height $H = 1.3$ m and period $T = 3$ s. Assume that $C_d = 0.8$ and $C_m = 1.8$. ■ Assuming that $C_L = 1.2$, estimate also the maximum transverse force per unit length acting on a section of the pile near the SWL and its predominant frequency.

8.6 A surface-piercing vertical circular caisson of diameter $D = 10$ m in water of depth $d = 10$ m is subjected to an incident wave train with height $H = 2.0$ m and period $T = 5.0$ s. What is the maximum overturning moment acting on the caisson about its base? ■ At what time, in seconds before a wave crest crosses the vertical axis, is the overturning moment a maximum? What is the maximum runup around the caisson? *[For the last part, you should rely on Reference Solution A9 in Appendix A.]*

8.7 ■ *[Extended problem – suitable as a paper assignment.]* Information on the tangential flow velocity u_θ around the base of a large circular caisson subjected to an incident wave train is needed to assess scour. u_θ may be obtained by differentiating the velocity potential ϕ using

$$u_\theta = \left(\frac{1}{r}\right)\frac{\partial \phi}{\partial \theta}$$

and then taking $s = 0$ and $r = a$. By considering the expression for velocity potential given in Table 8.2, and using the notation in the table, develop an expression for $u_{\theta o}/u_o$, in terms of ka and θ, where $u_{\theta o}$ is the amplitude of the tangential velocity around the base of the cylinder ($s = 0$, $r = a$) and u_o is the amplitude of the velocity at the seabed in the absence of the cylinder. In undertaking this derivation, you will need to rely on the following identity:

$$J_m(ka) - \frac{J'_m(ka)}{H_m^{(1)\prime}(ka)} H_m^{(1)}(ka) = \frac{2i}{\pi ka H_m^{(1)\prime}(ka)}$$

By comparing the expression you have developed with that for the runup around the cylinder given in Table 8.2, adapt the spreadsheet solution for wave runup given in Reference Solution A9 in Appendix A so as to develop a spreadsheet solution for $u_{\theta o}/u_o$ for a range of values of θ and a given value of ka. Thereby plot the distribution of $u_{\theta o}/u_o$ around the caisson base for the case $ka = 1.0$. What are the maximum values of $u_{\theta o}/u_o$ for $ka = 0.1$ and 1.0 and at what values of θ do they occur?

8.8 A rectangular-section floating breakwater has a beam $B = 5$ m, a draft $h = 2$ m, and a freeboard $f = 0.5$ m and it is located in water of depth $d = 8$ m. Assuming that the breakwater is fixed, develop a plot of K_t versus B/L on the basis of the Macagno formula. Estimate the heave and roll natural periods. Below what incident wave period is the fixed-breakwater assumption reasonable, assuming that this should not be within 10% of a resonant period? Estimate the transmission coefficient for an incident wave period $T = 3.2$ s.

8.9 ■ A rectangular-section floating dock of beam $B = 4$ m and draft $h = 1$ m in deep water is constrained by piles so as to undergo heave motions only. The incident waves correspond to $H = 0.8$ m and

$T = 3$ s. A computer model predicts that, for this wave period, the heave added mass μ_2 is given by $\mu_2/\rho Bh = 0.5$, and the heave damping coefficient λ_2 is given by $\lambda_2/\rho\omega Bh = 0.6$, where ω is the wave angular frequency. By considering the relevant equation of motion for heave, estimate the heave natural period and heave amplitude. [*You will need to rely on the Haskind relations.*]

Lagoons. These (Figure 9.1g) are elongated bays that tend to be parallel to the coast and are protected from the open ocean by barrier islands or spits. They typically have high salinity due to an excess of evaporation over precipitation, with little influx of freshwater and limited tidal exchange because of the barrier islands or spits.

Barrier islands. These (Figure 9.1h) are thin strips of land parallel to the shoreline, typically fronting lagoons or estuaries. They include beaches, adjacent dunes, and other environments. They form from the combined action of wind, waves, and longshore currents where sediment supplies are abundant and wave and tidal energy are conducive to onshore sand accumulation.

Spits. A spit (Figure 9.1i) is an extended stretch of land, often a beach, joined to the predominant coastline at one end and projecting into the sea at the other end. Spits are generally associated with sediment deposition over an extended time, arising from littoral drift that continues into an area beyond which the coastline bends landwards. In some cases, a spit may front a lagoon (see Figure 9.1g).

Coastal wetlands. These (Figure 9.1j,k) are coastal areas that are permanently or seasonally saturated with seawater and contain aquatic vegetation. They may be present in low-lying protected areas, such as the intertidal zone of an estuary, or on the landward side of barrier islands. They can include *salt marches*, which have rich, waterlogged soils that support various salt-tolerant plants and grasses, and *swamps*, which are predominantly forested and include *mangrove swamps* with mangrove trees in tropical and subtropical zones (Figure 9.1k).

Coastal bluffs, rocky coasts, and glaciated coasts. Coastal bluffs (Figure 9.1l), which may be either vegetated or bare, may erode intermittently over the long term due to extreme storms and tides, coupled with slides and mass wasting. Rocky coasts (Figures 9.1m,n) may extend over a wide area or be localized as headlands. They may be composed of less erodible materials, featuring cliffs, arches, and caves (Figure 9.1m), or they may be more prone to erosion (Figure 9.1n). Glaciated coasts (Figure 9.1o), of which fjords are a well-known example, are associated with active or former glacier activity, as modified by coastal processes. Typically, they have steep slopes, with abundant rock outcrops.

Reef coasts. Finally, reef coasts (Figure 9.1p) are "mounds" built up from the seabed by the interaction of organisms and coastal processes, usually with a skeletal framework formed by the organisms. Coral reefs, formed with corals and calcareous algae, are the most common. Reef configurations may include a fringing reef (attached to an island), a barrier reef (around an island, resulting in a lagoon), and an atoll, which is a ring reef with little or no land present (as in Figure 9.1p).

9.3 Sediment Properties

Sediment properties relevant to an assessment of coastal processes are summarized.

9.3.1 Sediment Size

Various schemes are used to classify sediment size, with Table 9.1 providing a common classification scheme derived from what is referred to as the Wentworth grain size classification.

In classifying sediment sizes, various definitions of particle diameter are possible, with "sieve size" commonly used. This corresponds to the size of sieve openings used to measure sediment size distributions. Some variations in terminology are used (e.g. coarse gravels may be referred to as pebbles or

Table 9.1 Sediment size classification.

Category	Particle diameter (mm)
Boulders	>256
Cobbles	64–256
Gravel – coarse	16–64
Gravel – medium	8–16
Gravel – fine	2–8
Sand – coarse	0.5–2
Sand – medium	0.25–0.5
Sand – fine	0.063–0.25
Silt	0.004–0.063
Clay	<0.004

shingle), and additional degrees of classification (e.g. coarse, medium, and fine silts) are used. Clays are also distinguished by cohesivity and plasticity.

9.3.2 Cohesive Sediments

Cohesion refers to the tendency of sediment particles smaller than about 0.005 mm to attach to each other because of attractive electrostatic charges. Therefore, particles in suspension collect into "flocs" that are loosely clumped aggregates and so have a much higher fall velocity than do the individual particles. Cohesive sediments on the seabed are more resistant to motion and transport than are noncohesive sediments. Mud refers to a cohesive sediment composed of a mixture of finer silt and clay, with some organic matter. Muds are typically found in sheltered areas protected from strong wave and current activity, such as the upper and mid reaches of estuaries and lagoons. The behavior and transport of cohesive materials differ from those of sand and other noncohesive materials and are discussed in Section 9.6.6.

9.3.3 Sediment Composition and Density

The most common constituent of sand is silica (silicon dioxide), usually in the form of quartz. The sediment density ρ_s, sometimes referred to as grain density in order to distinguish this from the bulk density described below, is often taken as 2650 kg/m^3. The density of seawater, denoted ρ, is taken as 1065 kg/m^3, so that the submerged specific gravity $s' = \rho_s/\rho - 1$ has a value of 1.49. For river flows, the density of freshwater, denoted ρ_f, is taken as 1000 kg/m^3, so that the submerged specific gravity is then $s'_f = \rho_s/\rho_f - 1 = 1.65$. Coastal sediments may include, for instance, eroded limestone, coral and shell fragments, and volcanic basalts. Such changes in composition may affect the density.

9.3.4 Porosity and Bulk Density

The porosity n of a sediment sample is defined as the ratio of the volume of voids to the total volume. Voids are the spaces in the sample that are occupied by water or air. Typical values of n are as follows:

Gravel: 0.24–0.38

Coarse sand: 0.31–0.46
Fine sand: 0.26–0.53

Bulk density refers to the mass of a sediment sample per unit volume, recognizing that the sample includes solids, water, and air. The bulk density depends on porosity and the extent to which voids are filled by air or water. The dry bulk density is given as $n\rho_s$, while the saturated bulk density is given as $n\rho_s + (1-n)\rho$.

9.3.5 Fall Velocity

The fall velocity, also termed the settling velocity, refers to the terminal velocity of a sediment grain falling in a stationary fluid. This is an important property that influences sediment transport characteristics. The fall velocity may be estimated by balancing the submerged weight of a sediment particle against the drag force acting on the particle. A force balance leads to

$$(\rho_s - 1)g\left(\frac{\pi D^3}{6}\right) - \frac{1}{2}\rho w^2 C_d\left(\frac{\pi D^2}{4}\right) = 0$$

where w is the fall velocity, D is the effective diameter of the particle, ρ_s is the particle density, and C_d is the drag coefficient. Using $s' = (\rho_s - \rho)/\rho$ as the submerged specific gravity, an expression for the fall velocity can be obtained as

$$w = \sqrt{\frac{4s'gD}{3C_d}}$$

The drag coefficient C_d varies with Reynolds number $\text{Re} = wD/\nu$, where ν is the kinematic viscosity of water. For small particles corresponding to a laminar flow, $C_d = 18/\text{Re}$, while for larger particles corresponding to a turbulent flow C_d is a constant. Since particles are not spherical, C_d also depends on a shape factor.

Various studies have been undertaken to enable the fall velocity w to be determined for a range of conditions. Ferguson and Church (2004) have proposed a simplified equation for w that conforms to the laminar and turbulent limits for small and large values of D, respectively, and that is dimensionally homogeneous. This is given as

$$w = \frac{s'gD^2}{C_1\nu + \sqrt{0.75C_2 s'gD^3}}$$

When account is taken of the shape factor for typical grains and when D refers to diameter based on sieve size, the constants are given as $C_1 = 18$, $C_2 = 1.0$.

As may be inferred from the summary of cohesive sediments provided earlier, the above description of fall velocity does not relate to cohesive sediments. Since the effective diameter of the flocs is greater than that of the individual sediment particles, the fall velocity is accordingly greater than that based on particle size and is notably influenced by suspended sediment concentration. At lower concentrations, the fall velocity increases with increasing concentration, since increasing concentration tends to increase the probability of floc formation and floc size. At very high concentrations, however, neighboring flocs are in close proximity to each other. This interference tends to cause a decrease in the settling velocity with increasing concentration, referred to as "hindered settling." Formulations are available for estimating the fall velocity of cohesive sediments in terms of the particle/floc size and the volume concentration of suspended matter.

9.4 Threshold of Sediment Motion

9.4.1 Unidirectional Flow

For a unidirectional flow over a sediment bed, the sediment is set in motion by virtue of the shear stress acting on sediment particles due to the fluid flow giving rise to a sufficiently large overturning moment. Because of this, the condition for the onset of sediment is usually expressed in terms of a critical shear stress at the seabed, denoted τ_w. τ_w is commonly expressed in the form of a friction velocity u_* that is defined as $u_* = \sqrt{\tau_w/\rho}$.

The friction velocity corresponding to the threshold for sediment motion depends on the grain size D, the submerged specific gravity s', the fluid viscosity ν, and the gravitational constant g. On the basis of a dimensional analysis, u_* may be expressed in the form:

$$\frac{u_*^2}{s'gD} = f(Re_*)$$

$u_*^2/s'gD$ is termed the Shields parameter and $Re_* = u_*D/\nu$ is a Reynolds number based on the grain diameter and friction velocity, referred to as a grain size Reynolds number. Originating from work by Shields in 1936, an approximation to this relationship is indicated in Figure 9.2.

It is emphasized that the curve shown in Figure 9.2 is approximate, recognizing that measurements exhibit notable scatter and that there may be different interpretations as to the precise condition for incipient sediment motion. Even so, the curve corresponds to a laminar boundary layer at sufficiently low values of Re_*, a turbulent boundary layer at sufficiently high values of Re_*, and a transition region in between.

The figure includes broken lines corresponding to approximations for the limiting cases of small and large values of Re_*, where the following simplification may be used.

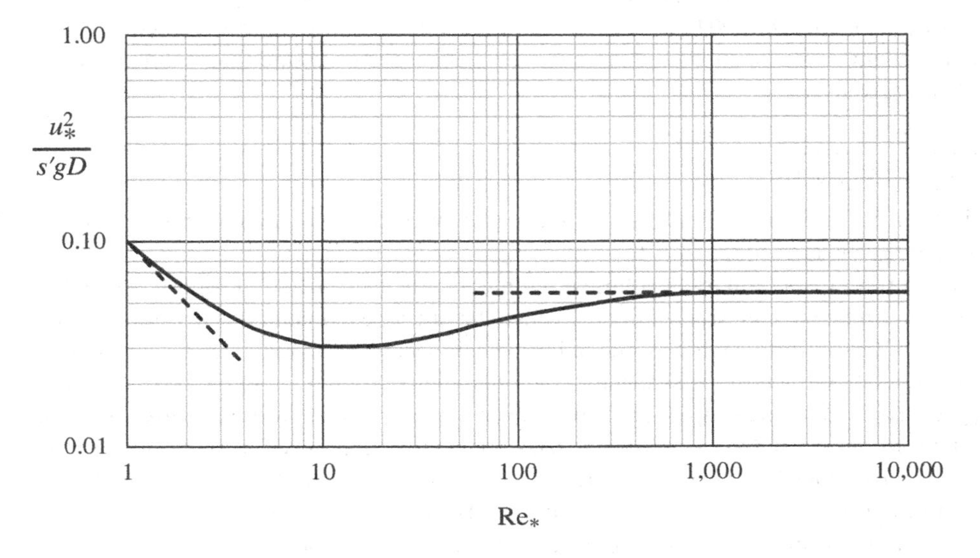

Figure 9.2 Shields parameter as a function of Reynolds number Re_*.

$$\frac{u_*^2}{s'gD} = \begin{cases} 0.1/\mathrm{Re}_* & \text{for } \mathrm{Re}_* < 5 \\ 0.056 & \text{for } \mathrm{Re}_* > 70 \end{cases}$$

Note that Reynolds number ranges shown above are less restrictive than is apparent in Figure 9.2.

In order to determine the conditions for the onset of sediment motion more readily, there is a need to express τ_w in terms of the prevailing flow conditions, usually described by the depth-averaged flow velocity U. Such relationships are described in the literature relating to open channel flows, whereby τ_w is related to U in terms of the roughness of the seabed or river bottom. Alternative approaches to developing such a relationship refer to a friction coefficient, Chezy coefficient, or Manning n value. With respect to the latter, the average velocity U is given in terms of the Manning n value for a wide river as

$$U = \frac{d^{2/3}S^{1/2}}{n} \quad \text{(SI units)}$$

where d is the water depth and S is the effective slope of the water surface. Also, τ_w is related to the surface slope S by $\tau_w = \rho g d S$. This leads to

$$U = \frac{1}{n}\sqrt{\frac{\tau_w d^{\frac{1}{3}}}{\rho g}} \quad \text{(SI units)}$$

Example 9.1 ***Incipient Sediment Motion due to a Current***

What current speed will cause sediment of size $D = 0.25$ mm to be set in motion at a coastal location with no waves and a depth of 3 m? Assume a Manning n value of 0.03.

Solution

Specified parameters

$g = 9.80665\ \mathrm{m/s^2}$
$\rho = 1025\ \mathrm{kg/m^3}$
$\rho_s = 2650\ \mathrm{kg/m^3}$
$\nu = 1.0 \times 10^{-6}\ \mathrm{m^2/s}$
$d = 3.0\ \mathrm{m}$
$D = 0.25\ \mathrm{mm}$
Manning $n = 0.03$

Assume turbulent conditions, Re$_*$ >70

$$s' = (\rho_s - \rho)/\rho) = 1.59$$

u_* for onset of sediment motion is given by

$$u_*/s'gD = 0.05$$

$$u_* = 0.05s'gD = 0014\ \mathrm{m/s}$$

Check Re$_*$ to validate the assumption:

$\mathrm{Re}_* = u_* D/\nu = 3.5$
$\mathrm{Re}_* < 70$ – hence the assumption is **incorrect**

Assume laminar conditions, Re* < 5

u_* for onset of sediment motion is given by $u_*^2/s'gD = 0.1/\text{Re}_*$

Hence:

$$u_* = (0.1\nu s'g)^{1/3} = 0012 \text{ m/s}$$

Check Re* to validate the assumption:

$\text{Re}_* = u_*D/\nu = 2.90$

$\text{Re}_* < 5$ – hence the assumption is **correct**

Therefore, use:

$$u_* = 0012 \text{ m/s}$$

Using $u_* = \sqrt{\tau_w/\rho}$

$$\tau_w = 0.14 \text{ Pa}$$

Now use the formula for U in terms of τ_w and the Manning n:

$$U = \frac{1}{n}\sqrt{\frac{\tau_w d^{1/3}}{\rho g}} = 0.15 \text{ m/s}$$

9.4.2 Waves

In the case of waves, the onset of sediment motion is now dependent on the oscillatory velocity near the seabed, which may itself be obtained from linear wave theory. Many formulae have been proposed and have been found to lead to widely differing estimates of the critical flow velocity, as described by Sleath (1984). Three approaches are outlined here.

An early formula, developed by Bagnold in 1946, provides the critical velocity as

$$u_o = 2.38\, s'^{2/3} D^{0.433} T^{1/3} \quad \text{(SI units)}$$

where u_o is the velocity amplitude at the seabed and T is the wave period. A second formula that has been widely used, as proposed by Komar and Miller in 1974, is

$$\frac{u_o^2}{s'gD} = \begin{cases} 0.21\,(d_o/D)^{1/2} & \text{for } D < 0.05 \text{ cm} \\ 0.46\pi\,(d_o/D)^{1/4} & \text{for } D > 0.05 \text{ cm} \end{cases}$$

where $d_o = 2u_o/\omega$ is the orbital diameter of the fluid motion near the seabed and ω is the wave angular velocity.

■➔

Another approach is based on extending the steady-flow Shields parameter methodology using the same relationship between the Shields parameter $u_*^2/s'gD$ and $\text{Re}_* = u_*D/\nu$, as provided in Figure 9.2. In applying this approach to a wave flow, u* is once more defined as $u_* = \sqrt{\tau_w/\rho}$, but τ_w is now taken as the amplitude of the shear stress on the seabed. In order to utilize the Shields relationship, u_* needs to be obtained in terms of the maximum flow velocity at the seabed, denoted u_o. This is usually developed by expressing u_* in terms of a friction factor f_w using $f_w = 2(u_*/u_o)^2$, where f_w is defined as $f_w = \tau_w/\frac{1}{2}\rho u_o^2$. Information on the friction factor was provided in Section 3.7.2. This included equations relating f_w to the wave Reynolds number $\mathrm{Re}_w = u_o a/\nu$ and the relative roughness parameter a/k_s, where a is the orbital amplitude of the fluid displacement at the seabed and k_s is a characteristic roughness size.

An estimate of the friction factor may be suitably combined with the Shields condition for the onset of sediment motion in order to establish conditions for the onset of sediment motion in waves. One of the problems in this chapter enables the reader to explore this approach for a particular case. However, while this may be undertaken for the smooth, turbulent and rough, turbulent flow regimes, this cannot be done for a laminar flow, i.e. when D is sufficiently small, since u_* does not then depend on D.

Example 9.2 ***Incipient Sediment Motion due to Waves***
Waves with a period $T = 4$ s approach a beach composed of sediment with size $D = 0.1$ mm. On the basis of the Bagnold and the Komar and Millar formulae in turn, estimate the wave-induced flow velocity amplitude at the seabed that is needed to set the sediment in motion.

Solution

Specified parameters

$g = 9.80665$ m/s^2
$\rho = 1025$ kg/m^3
$\rho_s = 2650$ kg/m^3
$T = 4.0$ s
$D = 0.1$ mm

Bagnold Formula for u_o

Bagnold formula is

$$u_o = 2.38\, s'^{2/3}\, D^{0.433}\, T^{1/3} \quad \text{(SI units)}$$

Therefore.

$u_o = 0.10$ m/s

Komar and Miller formula for u_o

The Komar and Miller formula indicates that, for $D < 0.5$ mm, incipient motion occurs when

$$\frac{u_o^2}{s'gD} = 0.21\sqrt{\frac{2U}{\omega D}}$$

Rearrange this to provide an expression for u_o:

$$u_o = [0.21\, s'g\, \sqrt{TD/\pi}\,]^{2/3}$$

Therefore.

$u_o = 0.11$ m/s

9.5 Beach Characteristics

Prior to a description of key characteristics of beaches, the terminology of various zones relevant to descriptions of a beach is summarized. The *backshore* refers to the area of the shore lying between the high tide level and the highest elevation affected by severe storms. The *foreshore* refers to the zone between the lowest and highest tide levels. The *nearshore* refers to the zone extending seaward from the low tide level to just beyond the location at which wave breaking occurs. Finally, the *offshore* refers to the zone extending seaward beyond the nearshore zone.

Beaches may differ with respect to material composition, material size, the wave/current/tidal climate, beach slope, and their horizontal extent, both parallel to and perpendicular to the shoreline. While sand beaches are predominant, beaches may also consist of rocks, gravels, cobbles, pebbles, and shingle.

Figure 9.3 Beach categories with respect to materials, steepness, width, and length. (a) Shingle beach – Gibsons, BC, (b) rocky beach – Gibsons, BC, (c) Black Sand Beach – Punalu'u, HI, (d) well-sorted beach – Murlough Beach, UK, (e) gently sloping beach – Porthdinllaen Bay, UK, (f) steep beach – Haleiwa, HI, (g) wide, long beach – Lakes Entrance, Australia, (h) bounded beach – Bondi Beach, Australia. *Sources:* (c) Diego Delso/Wikimedia Commons/CC BY SA 4.0. (d) Eric Jones/Wikimedia Commons/CC BY SA 2.0. (e) Eric Jones/Wikimedia Commons/CC BY SA 2.0. (f) Christopher Michel/Wikimedia Commons/CC BY 2.0. (g) Pimlico27/Wikimedia Commons/CC BY SA 4.0. (h) David Stanley/Wikimedia Commons/CC BY 3.0.

(e) (f) (g) (h)

Figure 9.3 *(Continued)*

A pocket beach refers to a small beach isolated between two headlands, where there is little or no exchange of sediments with adjacent shorelines. A barred beach refers to a beach with a nonuniform slope corresponding to the presence of one or more nearshore bars parallel to the shoreline.

Examples of beaches with different characteristics with respect to materials, steepness, width, and length are now indicated. Thus, Figure 9.1a, shown earlier, and Figure 9.3a–d show beaches with different materials, respectively indicating a sand beach, a shingle beach, a rocky beach, a black sand beach (formed from lava), and finally a well-sorted beach composed of a sand lower beach and a gravel berm. Figure 9.3e,f indicate differences in beach slope, comparing a gently sloping beach with a steep beach. Finally, Figure 9.3g,h refer to the horizontal extent of a beach, comparing a long, wide beach with one bounded within a bay.

9.6 Sediment Transport Processes

Important aspects of coastal engineering relate to an understanding of sediment transport behavior in a coastal environment, estimating the extent of sediment transport, erosion, and deposition that occurs, and undertaking related mitigation measures. *Erosion* refers to the removal of sediments causing a lowering of a beach or seabed, while *deposition* refers to the accumulation of sediments causing a raising of a beach or seabed. *Accretion* is often used synonymously with deposition, but more specifically refers to the build-up or net increase in beach or seabed elevation, occurring as a balance of erosion, deposition, and compaction processes.

As a general summary, sediment transport along a shoreline includes *onshore–offshore transport* (also referred to as *cross-shore transport*) and *longshore transport*. Offshore transport takes place when storms occur at high tides transporting beach materials offshore, while onshore transport takes place under more modest conditions that drive sediment shoreward so as to rebuild a beach. Longshore transport refers to sediment movement parallel to the shore and takes place when waves approach a shoreline obliquely. As well, contingent on the availability of sediment, intermittent erosion of the backshore may occur so as to provide a sediment source for the beach. A large portion of such erosion is usually attributable to a few more severe events each year.

Sediment may be moved as *bedload*, which refers to sediment rolling or sliding along the seabed, and/or as *suspended load*, whereby sediments are put into suspension and carried with the prevailing current.

Approaches relating to sediment transport at a site include: developing an understanding and a description of the coastal processes that are occurring and assessing their future evolution; developing estimates of sediment transport rates based on available information relating to sediment characteristics, wave and current climate, and bathymetry; and developing mitigation measures that alter natural sediment transport rates. Component aspects may include a study of historical photographs to examine past changes in shoreline evolution and surveys to obtain a series of beach profiles at selected transects along a beach and at selected times.

9.6.1 Onshore–Offshore Transport

Figure 9.4 illustrates the consequences of onshore–offshore transport on sand beaches. A "winter profile" is associated with storm activity as sediment is eroded from the beach and deposited as an offshore bar near the breaker line. A "summer profile" is developed under relatively calm, low swell conditions, allowing long period waves to transport sediment from offshore to repair and rebuild the beach so as to form a beach berm.

In the above summary, an offshore bar refers to a larger bedform that develops parallel to the shoreline as indicated in the figure. More generally, *bedforms* refer to patterns in the sediment along shorelines and in estuaries and rivers that include ripples, dunes, sand waves, and bars, resulting from the interactions of sediment with currents and waves. A beach berm, which may be referred to simply as a berm, is a ridge situated between the upper foreshore and backshore, that is usually composed of sands or gravels, and that is formed on account of the landward transport of coarser beach materials.

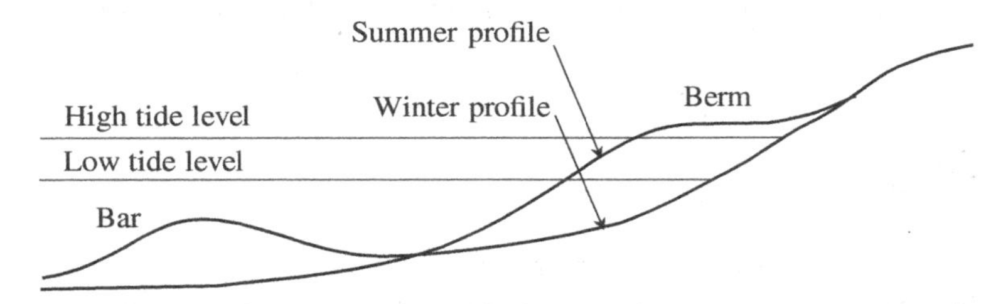

Figure 9.4 Illustration of summer and winter beach profiles.

Onshore–offshore transport is characterized by a nondimensional fall velocity parameter F, defined as $F = H_{sb}/wT$, where H_{sb} is the significant wave height at the breaker line, w is the median sediment fall velocity, and T is the wave period. Offshore transport occurs when $F > 1$ and onshore transport occurs when $F < 1$. While the magnitude of net sediment transport onshore or offshore is difficult to estimate, the volume of sediment moved offshore in a season is generally an order of magnitude less than the volume of longshore transport over the same period.

For locations with a high tidal range, the profile is further complicated, with beach profile changes extending significantly below the low tide level. For cobble beaches, changes in beach profile occur primarily on the upper portion of beach and less so below the mean water level.

9.6.2 Longshore Transport

Longshore transport, also termed *littoral drift*, refers to the movement of sediment parallel to a beach, associated with obliquely approaching waves and a resulting longshore current parallel to the beach, such that sediment is set into motion by the waves and then transported by the longshore current. Figure 9.5 illustrates water particle trajectories leading to the longshore current as waves approach a beach obliquely and the longshore current profile extending across the surf zone.

The longshore current magnitude varies with distance from the shoreline as indicated in the figure: it is zero at the shoreline itself, increases with distance from the shore to a maximum shoreward of the breaker line, and then reduces to zero at some distance beyond the breaker line. For a straight shoreline, the average longshore current across the surf zone, $\overline{V}$, is given approximately as

$$\overline{V} = 0.675\sqrt{gH_b}\,\sin(2\alpha_b)$$

where H_b is the breaking wave height and α_b is the angle between the wave crests and bottom contours at the breaker line.

For a long, straight beach, this current leads to the continuous movement of sand along the beach face. More generally, erosion occurs in areas of more intense wave activity and deposition occurs in areas of milder wave activity. A jetty, which acts as a partial barrier to longshore transport, causes deposition on one side and erosion on the other side.

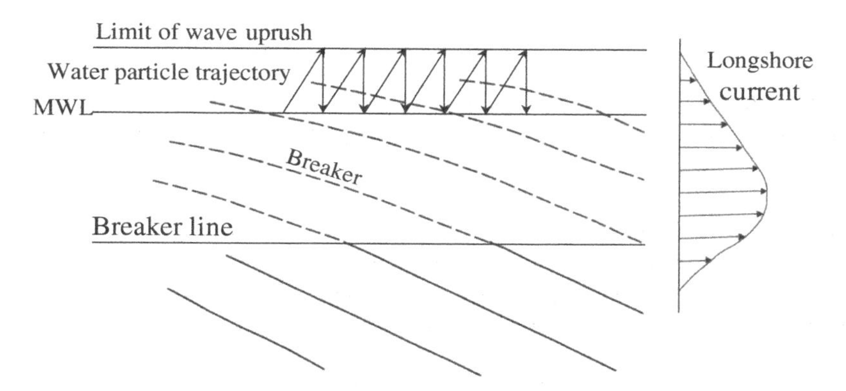

Figure 9.5 Illustration of longshore current due to oblique waves approaching a shoreline.

9.6.3 Estimates of Longshore Transport

Extensive work has been undertaken to develop methods for estimating the longshore sediment transport rate in terms of wave and beach characteristics. The most widely used approach, referred to as the "CERC formula" and described in the Coastal Engineering Manual (2002), is to rely on an expression for what has been referred to as the longshore wave energy flux at the breaker line, denoted P_ℓ, and then obtain the longshore transport rate Q (in units of m^3/s) as proportional to P_ℓ.

An expression for P_ℓ has been developed by applying the following expression for regular waves at the breaker line:

$$P_\ell = E\, c_G \, \sin(\alpha) \cos(\alpha)$$

where E is the wave energy density for regular waves, c_G is the group velocity, and α is the angle between the wave crests and the seabed contours. This may be used to develop an expression for P_ℓ in terms of H_{sb}, the significant wave height at the breaker line, and α_b, the angle between the wave crests and the seabed contours at the breaker line ($\alpha_b = 0$ for wave crests parallel to the beach).

However, there are two caveats to doing so. First, it has been pointed out that the above expression for regular waves is more appropriately derived with respect to the longshore component of momentum flux, related to the radiation stress, and not the energy flux. However, this does not detract from its use, but rather from the "energy flux" descriptor that is more commonly used. Second, in extending the above formula to random waves, the regular wave height H would normally be replaced by the root-mean-square wave height H_r, whereas it is the significant wave height H_s that is more commonly used. While both forms of height have been used in different studies, the use of one relative to the other solely affects the empirical constant that relates Q to P_ℓ.

Bearing in mind these two caveats, the most common expression for P_ℓ in terms of H_{sb} and α_b is given as

$$P_\ell = \frac{1}{16\sqrt{\gamma}} \rho g H_{sb}^2 \sqrt{g H_{sb}} \, \sin(2\alpha_b)$$

where γ is the breaker index (i.e. the height-to-depth ratio at breaking).

When considering the sediment transport rate, it is the immersed weight transport rate, rather than the volume transport rate, that has units corresponding to those of P_ℓ. This is taken as proportional to P_ℓ with a dimensionless proportionality constant K. Recognizing this, the volume transport rate itself, Q (in units of m^3/s), is expressed in terms of P_ℓ as

$$Q = \frac{K}{(\rho_s - \rho)(1-n)g} P_\ell$$

where n is the porosity (ratio of the volume of voids to total volume) of the sediment, which may be taken as approximately 0.4 for a sand beach and the empirical coefficient K is approximately 0.4. Note that, if the root-mean-square wave height H_r is used in place of H_s in the formula for P_ℓ, then K would be twice this value, i.e. approximately 0.8.

It is noted that wave conditions, including relative direction, may vary along a beach, in part because of a changing beach orientation. This variation leads to differential sediment movement along the beach, with some portions eroding and some portions accreting. Thus, if the longshore sediment transport rate is constant, given sufficient sediment supply there will be no net erosion of the beach. If this rate increases in the direction of sediment movement, there will be net erosion, while if it decreases, there will be net deposition. Also, if there is an insufficient supply of sediment for this natural process to continue, then beach erosion will occur.

bluff toe erodes when exposed to severe wave and storm activity, the bluff itself erodes as the toe erodes, placing sediments on the beach, and sediment transport on the beach removes eroded materials.

Bluff recession is highly episodic, both with respect to toe erosion due to waves in extreme storms occurring at high tides and with respect to subsequent bluff mass wasting, often associated with heavy rainfall, that leads to discrete slope failures separated by long periods of little or no change. As well, bluff recession is spatially variable, often over very short distances.

In summary, the mechanisms causing erosion are complex, recession rates exhibit significant temporal and spatial variations, and there is a high degree of uncertainty in predicting these. Even so, a range of models (e.g. Limber et al. 2018) have been developed to estimate long-term bluff recession rates. A key aspect of these is the need to account for the effects of future sea level rise. Most models are developed by expressing the rate of toe recession as proportional to a forcing function that accounts for relative sea level rise (RSLR), wave and water level conditions, beach conditions, and bluff conditions, with different models representing these in different ways.

One simplified recession model relies on the assumption that the current long-term recession rate continues indefinitely and is supplemented by an additional recession on account of RSLR. The latter value is estimated by assuming that the equilibrium beach profile remains unaltered over the long term, so that the toe of the bluff is shifted upward by the RSLR value and landward by an amount that is consistent with maintaining the equilibrium beach profile. However, in common with many of the models, this presumes that the bluff is fully responsive to toe erosion such that the entire bluff recedes along with the toe. This is expected to be unduly conservative. However, improved estimates may then be developed by taking account of a slope stability analysis that more reasonably accounts for the bluff's resistance to toe erosion.

9.8 Scour

Scour refers to the localized removal of sediment around the base of a structure due to wave and/or current flows. Scour has the potential to undermine structures and thereby lead to structural failure. In waves, scour occurs when the wave-induced oscillatory velocity adjacent to the seabed, amplified at certain locations by the presence of a structure, is sufficiently high to cause sediment movement, coupled with a current that moves sediment away. Fine sands and silts are more likely to be carried as suspended load, whereas coarser materials may be transported as bedload. Sumer and Fredsoe (2002) provide a detailed outline of the mechanics of scour and of scour predictions for a wide range of situations. Scour is a notable design constraint in coastal engineering works and may occur at jetties and piers, vertical and sloped seawalls, rubble-mound structures, pilings, submarine pipelines, breakwaters, and caissons.

9.8.1 Scour Depth Predictions

Predictions of scour depth are of interest and selected predictions of wave-induced scour are summarized below. (Current-induced scour is not considered.) Predictions of scour depth are difficult due to a large number of variables (including the structural configuration, wave conditions, and sediment properties), complex physical processes, difficulties in scaling from laboratory measurements, and the restricted applicability of field measurements.

Limited formulae and rules of thumb are available for certain situations in regard to different categories of structure (various kinds of piles, walls, and pipelines), different descriptions of wave conditions (regular and irregular waves, breaking and nonbreaking waves, normally incident and oblique waves), the extent of coexisting currents, and different ranges of sediment size. Here, a few miscellaneous formulae for the maximum scour depth, S_m, extracted from the Coastal Engineering Manual (2002), include the following:

- For normally incident, nonbreaking regular waves approaching a vertical wall:

$$\frac{S_m}{H} = \alpha\,[\sinh(kd)]^{-1.35}$$

where $\alpha = 0.4$ for fine sands and $\alpha = 0.3$ for coarse sands. The maximum scour tends to occur at a node of the associated standing wave train where seabed velocities are maximum.

- For oblique waves approaching a vertical wall, sediment movement occurs parallel to the wall, such that scour may occur and be more pronounced close to the base of the wall.
- For breaking waves at vertical walls, S_m is taken to correspond to the maximum nonbreaking wave height.
- For small diameter piles, $S_m/D \leq 2$, where D is the pile diameter. For this case S_m is notably influenced by the presence of a current.

For cohesive soils, methods are less well developed. The extent of scour depends on the shear stress and flow velocity at the seabed and on the in situ strength of the seabed material in resisting scour. This in turn depends on the soil properties.

9.8.2 Scour Protection

Approaches to scour mitigation or prevention generally rely on the placement of rock armor or other protective systems around a structure's foundation or in areas susceptible to scour. Types of protective systems include sandbags/geotextile bags, concrete mattresses (possibly anchored to the seabed), and flexible frond mats. Guidance on rock armor sizing is provided in the Rock Manual (2007).

9.9 Mitigation of Erosion and Accretion

Attention is now given to mitigation measures that can be undertaken to prevent or limit unwanted beach erosion or unwanted accretion of sediments. Mitigation measures applicable to scour are indicated in Section 9.8, while mitigation measures with respect to shoreline protection, so as to prevent or limit bluff or backshore erosion, are indicated in Section 9.10.

9.9.1 Beach Erosion

Examples of mitigation measures against unwanted erosion of beaches are illustrated in Figure 9.7. Figure 9.7a shows a groin field, Figure 9.7b shows a series of offshore breakwaters, Figure 9.7c shows rock clusters, and Figure 9.7d shows a beach replenishment scheme in progress.

Summary descriptions of various mitigation measures are given below.

Groins. These are long narrow structures made of wood, concrete, or rock piles and are built perpendicular to shore (Figure 9.7a). The intent is that these structures trap sediment or limit the littoral drift along a beach, thus allowing the beach to build up on the updrift side.

Figure 9.7 Types of erosion mitigation schemes. (a) Groin field – Rostock, Germany, (b) offshore breakwater system – Rimini, Italy, (c) rock mounds – Vancouver, BC, (d) beach replenishment – Chapel St. Leonards, UK. *Sources:* (a) Federal Waterways Engineering and Research Institute/Flickr/CC BY. (b) Google Earth. (c) ShoreZone/CC BY 3.0. (d) Mark Gadsby/CC BY SA 2.0.

Offshore breakwaters. These lead to lower energy in sheltered areas so that sediment can build up there. Figure 9.7b illustrates a series of offshore breakwaters that have given rise to this build-up in the longitudinal profile of the beach. In some cases, sediment deposition in the sheltered area may lead to the exposed beach extending up to the breakwater itself. Where circumstances permit, one approach is to develop these as a series of vegetated islets (e.g. mangrove islets) so as to be nature-based.

Rock formations. Rock groins, rock mounds, and rock clusters can be designed in various ways to develop, in effect, one or more small, sheltered bays that trap sediment or create low energy zones to promote sediment accumulation. Figure 9.7c illustrates a series of rock mounds that serve to trap sediments. A range of such designs may be adopted. A closely related approach is the suitable placement of rock mounds or groins (sometimes obliquely oriented) so as to cause, in effect, small pocket beaches between adjacent headlands.

Beach nourishment. Beach nourishment, also referred to as artificial nourishment, corresponds to the artificial replacement of eroded material, as illustrated in Figure 9.7d. A plan for beach nourishment normally describes the extent and frequency of replacement, the grain size of replacement materials, and the specific areas where materials are placed.

Mats and rock beds. In some cases, erosion is limited by the suitable placement of mats or coarser materials. This approach may be adopted along the side slopes and banks of a navigation channel.

9.9.2 Sediment Accretion

Mitigation measures against unwanted accretion, such as the silting up of harbors, may include the following:

Sediment barriers. A sediment barrier may be designed so as to prevent or minimize sediment accumulation in harbors and their entrance channels. More commonly, suitably located groins, jetties, breakwaters, and other coastal structures act as sediment barriers. Fabric curtains, referred to as silt or turbidity curtains, may be deployed on a temporary basis to prevent unwanted transport of suspended silts beyond the area of dredging or construction.

Dredging. Perhaps, the most common approach is to undertake dredging as may be necessary so as to remove sediment that has accreted within a harbor or a navigation channel.

Sand bypassing. A sediment bypassing scheme involves pumping sand in a slurry between locations, typically from the updrift side to the downdrift side of a harbor entrance or entrance channel. Such schemes are used when littoral drift is significant for much of the year, as is the case of long, well-developed sand beaches exposed to the open ocean.

9.9.3 Coastal Entrances

Coastal entrances to lagoons, tidal inlets, and rivers are common features of coastlines. Typically, natural processes deposit sediment at a river delta in the form of a bar across the river mouth, with dynamic processes associated with tides, currents, and waves acting to remove sediments from the bar. In cases where the coastal entrance is used for navigation, there is a need to ensure that channel widths and depths are sufficient and that they remain so. Most commonly, the entrance channel is dredged to a sufficient depth, in conjunction with the placement of jetties that stabilize the entrance channel with respect to sediment movement from littoral drift. The jetty system controls the channel geometry, prevents the channel from shifting, and limits sediment deposition from the updrift side to the river mouth, while tidal and river flows flush the river so as to maintain a sufficiently deep channel. Guidelines on jetty requirements regarding the channel width and depth are available (e.g. Coastal Engineering Manual 2002).

9.10 Approaches to Shoreline Protection

While the preceding section was focused on limiting beach erosion and addressing unwanted accretion, more general mitigation measures are needed for shoreline protection in order to minimize coastal recession and prevent coastal flooding. These include traditional methods, sometimes referred to as "hard" protection schemes, and nature-based schemes that promote habitats and minimize impacts on the environment. The latter are sometimes referred to as "soft" protection schemes. Combinations of both methods are referred to herein as hybrid protection schemes. Prior to summarizing these, the concept of coastal resilience, which is relevant with respect to nature-based schemes, is outlined below.

9.10.1 Coastal Resilience

Resilience refers to the ability to "bounce back," recover from or cope with a harmful circumstance. Thus, an initial definition of coastal resilience might be understood as the ability to recover from or cope with the effects of extreme storms or other coastal hazards. However, the meaning of coastal resilience has become broader. Thus, harmful circumstances are taken to refer to both extreme events, such as storms, and to more gradual trends such as sea level rise. Also, coastal resilience is considered in the broader context of social, economic, and ecological systems. With respect to the latter, it emphasizes coastal management approaches that prioritize the preservation and protection of coastal areas such as beaches, coastal dunes, and tidal wetlands, while supporting nature-based interventions whenever feasible.

In describing coastal resilience, one perspective is that the ability to cope is taken to include the ability to "resist", i.e. to prevent damage in the first place by prior interventions (such as through the construction of a robust shoreline protection scheme). However, the prevailing view appears to be that coastal resilience is focused on accommodation, adaptation, and recovery, while the ability to resist is deemphasized. Typically, then, a resilient solution is one that allows for a greater probability of damage during an extreme event, but this is offset by a greater tolerance of the consequences, usually through a reliance on remediation, repair, and rehabilitation.

As an example relating to a shoreline protection scheme, a traditional (hard) scheme may be more robust in resisting storms than a nature-based scheme, but may have negative impacts such as undue sediment erosion and habitat reduction. On the other hand, while a nature-based scheme may be less robust in preventing coastal erosion or flooding, it may be designed to be more resilient through modified usage of adjacent land, including increased setbacks, suitable landscaping, improved drainage, and/or remediation of affected areas after severe storms. Overall, a resilient, nature-based design may be more attractive, more economical, and more ecologically sound than reliance on a traditional method.

9.10.2 Traditional Methods

Traditional or "hard" shoreline protection schemes include seawalls, rock armor, revetments, and gabions. Examples of seawalls and rock armor were provided in Figure 8.1a,b, respectively, while examples of a revetment and a gabion are provided in Figure 9.8a,b, respectively. Summary descriptions of these are given below.

Seawalls. A seawall is constructed from concrete, steel, or stone and located along a shoreline, with an example provided in Figure 8.1a. A wide range of configurations are possible.

Rock armor. Rock armor, also referred to as rubble-mound protection or riprap, corresponds to boulders or rocks piled up along the backshore or in front of a seawall. An example is provided in Figure 8.1b. Chapter 8 provides an indication of the range of designs that are possible.

Revetments. A revetment (Figure 9.8a) is an inclined structure made from concrete, wood, or rocks placed along a steep backshore.

Gabions. A gabion (Figure 9.8b) comprises a series of metal mesh cages or boxes filled with soil or rocks located along the backshore.

These schemes serve to dissipate wave energy and protect the backshore or cliffs from erosion. In the case of schemes intended to prevent toe erosion of a bluff or cliff, this should be undertaken in conjunction with measures to promote slope stability and cliff stabilization.

(a) (b)

Figure 9.8 Selected shoreline protection schemes. (a) Revetment – Mundesley beach, UK, (b) gabion – Studland, UK. *Sources:* (a) Kolforn/Wikimedia Commons/CC BY SA 4.0. (b) N. Chadwick/CC BY SA 2.0.

9.10.3 Nature-Based and Hybrid Methods

Greater recognition of the need to preserve habitat and minimize impacts on the environment has resulted in the increased prominence given to nature-based methods of shoreline protection. However, for locations exposed to more severe storm conditions, these methods alone may provide insufficient protection from shoreline erosion and coastal flooding. As a result, they are most effectively utilized when the consequences of insufficient protection may be accommodated, or when undertaken in conjunction with aspects of traditional methods, referred to here as hybrid methods. Often, a suitable combination of these methods may be more robust in limiting shoreline erosion and flooding during the most extreme storms, while preserving habitat.

From the perspective of coastal defense design, nature-based schemes represent a preferred response to design criteria that give a greater weight to resilience, habitat enhancement, and minimizing environmental impacts and a lower weight to potential overtopping or failure under the most extreme conditions. Such shifts in criteria weighting may be associated with regulatory and permitting requirements or a required adherence to an Official Community Plan. The mix of criteria and their relative weighting depends also on an assessment of a site and of a project with respect to issues such as the project scale, the severity of storms, tidal range, land usage requirements, existing habitat, client requirements, public safety, and economic viability. Furthermore, a deliberate process of consultations and engagement with stakeholders is often needed in order to develop a suitable combination of traditional methods, which may be less likely to fail in extreme storms, and nature-based methods that provide improved support for marine habitat.

In the above context, different kinds of nature-based and hybrid schemes are now outlined, including those that refer to selected examples illustrated in Figure 9.9.

Nature-Based Methods

Dune stabilization and regeneration. When dunes form landward of the backshore, they act as a barrier to flooding. Schemes used to stabilize and regenerate dunes include fencing to direct pedestrian traffic, weeding, pest control, thatching, and planting, and the use of logs and dead trees. Figure 9.9a

(a) (b) (c) (d)

Figure 9.9 Selected nature-based and hybrid shoreline protection schemes. (a) Dune stabilization – Mrzezyno, Poland, (b) dynamic revetment – North Cove, WA, (c) hybrid scheme with riprap, a habitat bench, and native plantings – Richmond, BC, (d) schematic of a hybrid scheme with riprap, rock clusters, habitat benches, and native plantings. *Sources:* (a) Krzysztof Ziarnek Kenraiz/Wikimedia Commons/CC BY SA 4.0. (b) Cranberrydavid/Wikimedia Commons/CC BY SA 4.0.

provides an example of a dune stabilization scheme through plantings and fencing to limit pedestrian access.

Dynamic revetment. A dynamic revetment refers to an artificial berm, typically a cobble berm. Slope recontouring, in which portions of a beach profile are adjusted by the placement of materials, represents a variant of such a scheme. Figure 9.9b provides an example of an artificial cobble berm.

Beach nourishment. As described earlier, beach nourishment describes the periodic replacement of eroded sediment on the beach. This approach requires periodic maintenance and can be expensive.

Anchored drift logs. At locations where drift logs are common, one approach is to anchor selected logs, especially near the toe of a bluff, to act as a form of nature-based protection.

Native plantings. Native plantings may be undertaken to stabilize the sediment along the shoreline. In tropical and subtropical areas, mangrove plantings or mangrove restoration projects are widely used along a shoreline to dissipate wave energy, stabilize the soil, and prevent or limit erosion. Otherwise, plantings are usually insufficient alone and are used in conjunction with other protection methods.

Habitat bench. A habitat bench or intertidal bench provides a horizontal or low slope bench incorporated into a shoreline protection scheme. This is designed to promote habitat development through soil retention and the use of suitable plantings. In some cases, a natural bench may be suitably restored, modified, or adapted.

Other nature-based schemes include islets with plantings, especially mangrove islets, and rock clusters or rock mounds that may be suitably designed to reduce wave energy impacts and create low energy zones in order to promote the accumulation of sediments.

Some schemes include a deliberate reduction of protection levels. These include the following:

Increased setbacks. It may be possible to increase setbacks required for development, with or without shoreline defenses built near a setback line. Thus, a natural shoreline with coastal habitat is preserved seaward of a coastal defense and allows for occasional inundation from a storm with little or no consequence.

Managed retreat. Managed retreat refers to a deliberate approach that allows certain areas of the coast to erode and flood, rather than attempting to preserve the shoreline. Components of a managed retreat program may include buy-back or buy-out programs with respect to residential or industrial areas, designating no-build areas, green space development, habitat restoration in retreat-prone areas, and regulating the types of development adjacent to designated retreat areas.

Hybrid Methods

Hybrid methods, which contain elements of both traditional and nature-based methods, commonly include suitably designed riprap, supplemented by nature-based protection methods including habitat benches, native plantings, and rock clusters. Design variations may include the use of rounded boulders, lower slopes, and varying sectional profiles; the riprap may be interspersed with troughs or benches for soil, native plantings, and habitat development; and it may be fronted by irregularly placed rock clusters. As one example, Figure 9.9c shows low-slope riprap protection with native plantings, a lower elevation habitat bench, and additional riprap protection below the bench. Similarly, Figure 9.9d provides a schematic of a section that incorporates riprap with rounded boulders, habitat benches, native plantings, and rock clusters placed on the foreshore.

9.11 Coastal Restoration

Coastal restoration refers to the process of supporting the recovery of an ecosystem that has been degraded, damaged, or destroyed and promoting habitat protection and enhancement. Activities include redeveloping and landscaping shorelines, debris removal, contaminant removal, invasive species mitigation, shoreline cleanup, native plantings, and limiting pedestrian access, either temporarily or over the long term. Specific examples include the restoration of wetlands and salt marshes that provide productive habitats. Such activities are undertaken through a collaboration involving coastal engineers, environmental biologists, and landscape architects.

9.12 Coastal Management

A critical aspect of shoreline protection and mitigation measures relates to a need for the management and integration of such activities over a broader shoreline region. Thus, coastal management, or coastal zone management, refers to management activities with respect to this more extensive coastal zone.

Coastal management is often developed and implemented through a coastal management plan that sets out the principles, objectives, and management strategies that together assure that development activities in coastal areas suitably follow an overall land-use plan, relevant environmental and related policies are adhered to, and the approaches adopted are cost-effective and are socially and environmentally responsible. Such a plan may include usage designations (e.g. conservation, residential, industrial, and recreation areas) with guidelines, an implementation plan, and a monitoring plan (e.g. for the assessment of effective habitat restoration and enhancement initiatives).

A coastal management plan should identify the roles and responsibilities of different levels of government and other stakeholders. The plan may be incorporated into or be supplementary to an Official Community Plan for a municipality or other level of government. A more recent approach is to integrate such a plan into a municipality's traditional asset management plan, so that assets include all natural assets such as wetlands and natural shorelines, along with goals and performance indicators that incorporate nature conservation and environmental stewardship principles.

Coastal management plans usually incorporate two key principles. One is an integrated approach in that different portions of a coastline are interlinked with all coastline elements (land, water, people, and the economy) and are managed with one integrated strategy. Another is a sustainability requirement that is applicable to natural resources, economic development, employment, recreation, and community education.

Such plans may include a cost–benefit analysis in which anticipated costs of implementing various components are compared to the expected benefits, both tangible (i.e. monetary value) and intangible (e.g. habitat restoration and visual impacts), with benefits needing to outweigh costs in order for a component project to proceed.

Coastal management plans may also refer to implementation approaches, which include decision-making, consultations, communications, and engagement. Some aspects of these are considered in Chapter 11 as relating to the design of coastal infrastructure.

One approach to coastal management that is sometimes adopted refers to *adaptive management*. In recognition of uncertainties and an insufficient understanding of the consequences of coastal solutions, adaptive management is a long-term, iterative process that involves acquiring additional information regarding the coastal system and coastal responses and adapting solutions accordingly.

Problems

[Where relevant, you should rely on the values of physical constants provided in Appendix C.]

Assignments

9.1 At a coastal location with no notable wave activity, water depths range from 3 to 6 m and tidal currents range up to 0.8 m/s. Develop an expression for the maximum sediment size D that may be moved, in terms of the current U, the water depth d, the submerged specific density of the sediment, s', and the Manning n value, assuming that $\mathrm{Re}_* > 70$. If $n = 0.03$, what is the maximum sediment size that may be moved under the above conditions? What is the fall velocity of these sediment grains?

9.2 In a steady tidal flow, above what shear stress at the seabed will a fine sand with $D = 0.2$ mm undergo motion?

9.3 Using the Komar and Millar formula for finer sediment, develop an expression for the wave height H giving rise to incipient sediment motion for a given grain size D, wave period T, and depth parameter kd. Deep-water design wave conditions at a coastal location correspond to $H_o = 1$ m and $T = 4$ s and there is no wave refraction. The beach is made up of sand of median size $D = 0.3$ mm. Taking account of wave shoaling, estimate the water depth for incipient motion of the sand for this case. At this location, what is the orbital diameter of the water particle motions at the seabed?

9.4 ■ At a coastal location, the wave height $H = 1.2$ m, the wave period $T = 5.0$ s, and the water depth $d = 5.0$ m. Assuming that the flow over the seabed corresponds to a rough, turbulent boundary layer, estimate the maximum grain size on the seabed that will undergo motion. In order to do so, utilize the descriptions of the friction factor given in Section 3.7.2 and the Shields parameter given in Section 9.4.2 applied to a wave flow. Confirm that the conditions of a rough, turbulent boundary layer are met. [*You are advised to develop an iterative solution in which* D, a/k_s, f_w, u_*, D, … *are calculated in turn.*]

9.5 Waves approaching a beach composed of sediment of size $D = 0.5$ mm have a significant wave height at breaking $H_{sb} = 1.2$ m and a wave period $T = 6.0$ s. In which direction will any onshore–offshore transport occur?

9.6 Waves approaching a beach have a significant wave height at breaking $H_{sb} = 1.2$ m and the angle between the wave crests and the seabed contours at the breaker line $\alpha_b = 40°$. Subject to sufficient sediment supply, estimate the longshore sediment transport rate in units of m^3/hr.

Written Assignments

9.7 For a selected coastal location with a beach (possibly indicated by your instructor), review the characteristics of the site relevant to coastal processes, including adjacent fetches and shorelines using Google Earth, the bathymetry and other features using a hydrographic chart, representative tide elevations as indicated on a chart, tide tables, or a portal, and, if possible, the wave and current climate, beach profiles, and sediment sizes. If possible, this may be supplemented by a site visit or a set of photographs at the site. Discuss the features of the site with respect to active coastal zone processes relating to sediment erosion and accretion.

9.8 An intended residential development along a selected coastal location (possibly indicated by your instructor) has an eroding backshore. Discuss the factors relevant to selecting a potential shoreline protection scheme and develop a recommended process for selecting a suitable option.

10

Mixing Processes

10.1 Introduction

Some coastal engineering issues depend upon the extent to which pollutants or other substances are transported and mixed with surrounding waters. Example topics include a determination of the fate of a pollutant that is discharged into coastal waters, the transport of mud and fine sediments suspended in water, and water exchange and circulation patterns in harbors and estuaries. Other topics include salinity intrusion into rivers and the design of outfalls and diffusers affecting the extent and manner in which effluents are discharged into the marine environment. All these topics are central to the neighboring discipline of environmental fluid mechanics. Texts that relate to this discipline include Fischer et al. (1979), Rubin and Atkinson (2001), Shen et al. (2002), and Imberger (2012).

Water quality measures. Water quality refers to the chemical, physical, biological, and radiological characteristics of water. Water quality measurements are needed to assess the health of ecosystems. Indicators of water quality include salinity, temperature, dissolved solids, suspended solids, pH (acidity or alkalinity), and dissolved oxygen. Additionally, turbidity refers to an optical property that indicates the amounts of light that are scattered and absorbed by particles in the water, while total suspended solids (TSS) correspond to the mass of suspended solids per unit volume of water (in mg/l). Turbidity and TSS are related, although their relationship is somewhat unique to each individual location or situation. Finally, salinity describes the amount of dissolved salts in water and is usually measured in units of g/kg (i.e. grams of salt per kilogram of water) or, identically, ppt (parts per thousand).

The focus of this chapter is on estimating the concentration of a pollutant, contaminant, or tracer in the marine environment. (The term pollutant is used herein to represent any of these.) Thus, the primary variable of interest is pollutant concentration, c, which is the mass of pollutant per unit volume of solution, usually measured in units of g/m^3, or, identically, mg/L. In general, c is a function of time t and location so that, with reference to a coordinate system (x, y, z), this dependence may be expressed as follows:

- one-dimensional problems: $c(x, t)$
- two-dimensional problems: $c(x, y, t)$ or $c(x, z, t)$
- three-dimensional problems: $c(x, y, z, t)$

The primary mixing mechanism arises through diffusion such that the flux of pollutant results from a concentration gradient. In particular, Fick's law states that the mass of pollutant crossing a unit area

An Introduction to Coastal Engineering, First Edition. Michael Isaacson.

Companion website: www.wiley.com/go/coastalengineering

per unit time is proportional to the gradient in pollutant concentration, with the proportionality constant termed the *diffusivity* or *diffusion coefficient*, D. In one dimension, this is given as

$$q = -D\frac{\partial c}{\partial x}$$

where q is the mass flow rate of pollutant per unit area, c is the concentration of pollutant, and x is distance in the direction of the concentration gradient.

If a current is present, the pollutant is also carried with the current while undergoing diffusion. In this context, advection refers to the transport of pollutant by the prevailing current.

Molecular diffusion refers to the net transport of a pollutant due to molecular movement. However, in coastal engineering applications, the concept is extended to larger scale mixing arising from *turbulent diffusion* due to random, turbulent motions that occur at high Reynolds numbers, typically in response to shear flows. Turbulent diffusion occurs much more rapidly than molecular diffusion. In a similar way, *eddy diffusion* refers to diffusion due to motions arising from eddies, recognizing that these may vary widely in size, from microscales to geophysical scales. Nonuniformities in a current field, particularly a shear flow, may give rise to an even greater degree of mixing in the longitudinal direction, termed *longitudinal dispersion*. Various theoretical and experimental studies and field measurements relate to the diffusion and dispersion coefficients. A description of these coefficients is provided in Section 10.4.

10.2 Advection–Diffusion Equation

The advection–diffusion equation, which describes the advection and diffusion of a pollutant, may be developed as a governing equation that combines Fick's law with a continuity equation that reflects the conservation of pollutant mass, taking account of advection. This equation is developed below for one-, two-, and three-dimensional flows.

10.2.1 One-Dimensional Equation

In one dimension, the continuity equation is

$$\frac{\partial c}{\partial t} = -\frac{\partial q}{\partial x}$$

This is combined with Fick's law to yield a governing equation for c:

$$\frac{\partial c}{\partial t} = \frac{\partial}{\partial x}\left(D\frac{\partial c}{\partial x}\right)$$

If D is considered not to vary with x, then this reduces to

$$\frac{\partial c}{\partial t} = D\frac{\partial^2 c}{\partial x^2}$$

When advection is taken into account, the one-dimensional diffusion equation is extended to the one-dimensional advection–diffusion equation with D constant:

$$\frac{\partial c}{\partial t} + u\frac{\partial c}{\partial x} = D\frac{\partial^2 c}{\partial x^2}$$

Solution

Specified parameters

$U = 0.2\,\text{m/s}$
$m = 100\,\text{kg}$
$d = 5\,\text{m}$
$D = 0.5\,\text{m}^2/\text{s}$
$h = 100\,\text{m}$
$x = 500\,\text{m}$
For a two-dimensional flow, $M = m/d$ is used:
$M = 20\,\text{kg/m}$

Formula for $c(x, y, t)$

This problem corresponds to a solution to the advection–diffusion equation for a two-dimensional flow in the (x, y) plane with a point source at $(0, h)$ and a mirror source at $(0, -h)$, used to simulate the impermeable shoreline. Based on an extension to the two-dimensional solution given earlier, the solution is

$$c(x, y, t) = \frac{M}{4\pi Dt} \times \left\{ \exp\left[-\frac{(x - Ut)^2}{4Dt} - \frac{(y - h)^2}{4Dt}\right] + \exp\left[-\frac{(x - Ut)^2}{4Dt} - \frac{(y + h)^2}{4Dt}\right]\right\}$$

where M has units of kg/m for a two-dimensional flow. Along the shoreline $y = 0$ and so this reduces to

$$c(x, 0, t) = \frac{M}{2\pi Dt} \exp\left[-\frac{(x - Ut)^2}{4Dt} - \frac{h^2}{4Dt}\right]$$

The variation of c with time at $x = 500\,\text{m}$ may be developed in a spreadsheet and can be plotted as follows:

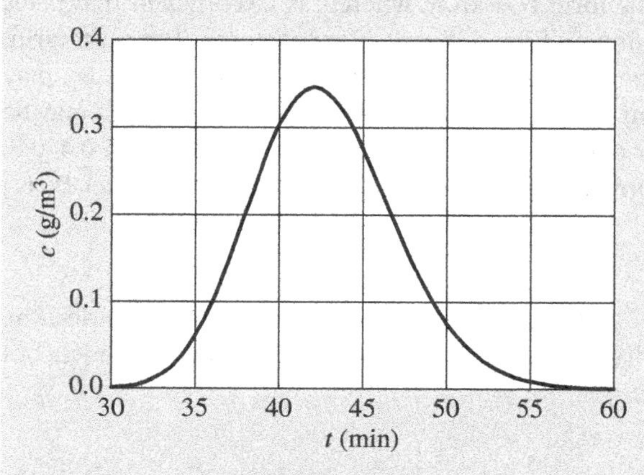

From the associated table, the maximum concentration c_m and the time t_m at which it occurs are obtained as

$c_\text{m} = 0.35\,\text{g/m}^3$
$t_\text{m} = 42\,\text{min}$

10.4 Diffusion and Dispersion Coefficients

The practical application of the various solutions to the advection–diffusion equation depends in part on knowledge of the diffusion or dispersion coefficients to be used. Prior to considering these, it is pertinent to elaborate on the mechanisms that cause different extents of mixing, often distinguished as vertical, transverse, and longitudinal mixing, which refer respectively to mixing in the vertical direction, the horizontal direction transverse to the flow, and the direction of flow. In this context, turbulent diffusion, which is associated with turbulent fluctuations in the flow, is generally not isotropic. Vertical mixing is related to the vertical velocity profile, which, in turn, is related to the bottom shear stress. Transverse mixing will be different in general, since there is no analogous bottom boundary with a related velocity profile.

However, the greatest difference occurs in the flow direction. While turbulent fluctuations in the longitudinal direction give rise to turbulent mixing in this direction, this turbulent diffusion is overshadowed by what is referred to as *longitudinal dispersion* associated with the vertical velocity profile. Specifically, the turbulent diffusion in the vertical direction combines with the shear flow arising from differential advection at different elevations to result in an effective diffusion in the longitudinal direction that is termed longitudinal dispersion. This is distinct from, and many times larger than, regular turbulent diffusion in the longitudinal direction.

Attention is now given to a selection of values for diffusion and dispersion coefficients. For reference, in the case of molecular diffusion, D for different substances is reasonably well known and is typically in the range $0.05–1.00 \times 10^{-4}\ \text{cm}^2/\text{s}$. In the cases of turbulent diffusion and longitudinal dispersion, approaches to estimating coefficients are quite complicated. Theories have been developed to relate D to the characteristics of the flow for various situations of increasing complexity. Many of these have been described in terms of the friction velocity u_* that is defined as $u_* = \sqrt{\tau_\text{w}/\rho}$, where τ_w is the shear stress at the seabed and ρ is the fluid density. Various measurements have been made and formulae proposed to enable D to be estimated under a wide range of circumstances.

The diffusion coefficient is often expressed in the form $D = ku_*d$, where k is a coefficient that takes on different values for the different forms of mixing and for different circumstances. For a uniform, straight, wide river of constant depth, the turbulent diffusion coefficient for transverse mixing, D_y, may be estimated by taking $k = 0.15$, while the turbulent diffusion coefficient for vertical mixing, D_z, may be estimated by taking $k = 0.07$. For natural, irregular rivers, transverse mixing corresponds to $k \approx 0.4–0.8$, while the longitudinal dispersion coefficient D_L corresponds to $k \approx 100–500$. One formulation for D_L is

$$D_\text{L} = 0.01\frac{U^2w^2}{u_*d}$$

where w is the river width, d is the average depth, and U is the average flow velocity. In estuaries, the situation is even more complicated. Longitudinal dispersion coefficients in estuaries can vary widely but are typically in the range $D_\text{L} \approx 100–300\ \text{m}^2/\text{s}$.

10.5 Stratified Flows

A stratified flow is a flow in which there is a variation in fluid density, usually as a result of salinity and/or temperature differences. As illustrative of vertically stratified flows, Figure 10.4 shows vertical distributions of salinity S for four examples of stratified flows (with the distributions shifted to the left or right so as not to overlap). Line (a) is referred to as a two-layer flow, which corresponds to two distinct layers of fluid with different salinities. Lines (b), (c), and (d) illustrate different examples of the interface between a river and an estuary corresponding to a highly stratified flow, a partially mixed flow, and a well-mixed flow, respectively. These cases are referred to in Section 10.6.

Topics of interest with respect to stratified flows include the behavior of long-period internal waves at the interface between two density layers, the outflow from a reservoir containing a stratified fluid, mixing in estuaries, salinity intrusion in rivers, and the behavior of jets and plumes. Of these, mixing in estuaries is described in Sections 10.6 and 10.7, salinity intrusion in rivers is summarized in Section 10.8, and the behavior of turbulent jets and plumes is summarized in Section 10.9. For further information on such flows, the reader is referred to texts in environmental fluid mechanics.

10.6 Mixing in Estuaries

An estuary refers to a partially enclosed transition zone between a river and the sea. Estuarine processes include hydrodynamics and circulation patterns, mixing, saltwater intrusion, and sediment movement. These various processes are complex and are influenced by a wide range of parameters, including the tidal range or tidal pattern, river flow rates, the complex bathymetry and shoreline profile, and sediment characteristics. A general outline of mixing in estuaries as influenced by stratification is now given. The reader is referred to Fischer et al. (1979) for a comprehensive description of this topic.

10.6.1 Categories of Estuaries

Estuaries may be categorized in various ways. A key approach is based on the degree of stratification in the region where freshwater from a river mixes with saltwater from the sea. Thus, a *highly stratified estuary* leads to a reasonably well-defined boundary between the saltwater and the freshwater from the river. A *partially mixed* or *partially stratified estuary* is one where saltwater and freshwater mix at all depths, but with a notable vertical gradient of salinity. A *well-mixed estuary* occurs when the river flow is low and tidally generated currents are moderate-to-strong, corresponding to a more modest vertical gradient of salinity. Corresponding salinity profiles are illustrated in Figure 10.3.

Estuaries may also be classified by their bathymetric and shoreline profiles. Thus, a fjord-type estuary contains a narrow sill adjacent to the open ocean, such that dense seawater seldom flows over the sill into the estuary. A coastal-plain estuary is relatively shallow with a more uniform depth, with tides providing an important source of mixing. Finally, bar-built and lagoon estuaries are relatively shallow with narrow mouths and sand bars at their entrances, often containing several channels, islands, and shoals.

10.6.2 Mixing Mechanisms

The primary mechanisms for mixing in estuaries are associated with the wind field, tidal flows, and river flow. First, winds blowing over the water surface result in shear stresses that in turn give rise to circulation patterns influenced by the bathymetry. Winds may or may not be a significant cause of mixing in

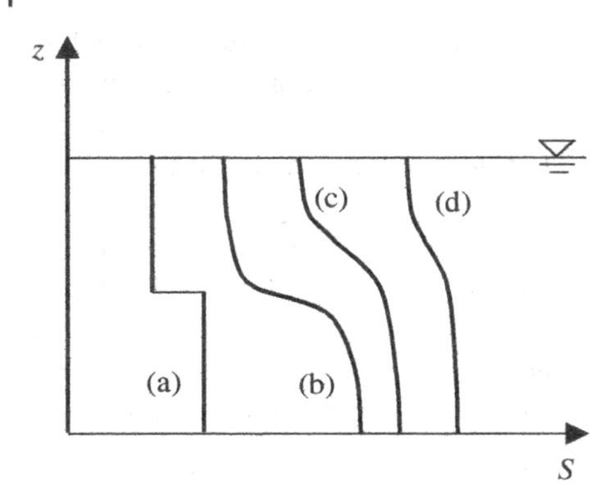

Figure 10.3 Representative vertical salinity distributions. (a) Two-layer flow, (b) highly stratified flow, (c) partially mixed flow, (d) well-mixed flow.

an estuary. Second, tidal flows give rise to various forms of mixing, including vertical mixing, transverse mixing, and longitudinal dispersion. These types of mixing are associated with complex, nonuniform bathymetric profiles, shear stresses along the seabed, vertical velocity profiles, and advection arising from circulation patterns. Third, a river discharging into an estuary leads to a stratified flow with various freshwater/saltwater interactions and associated mixing. The type of mixing includes the development of estuarine circulation driven by a longitudinal density gradient, with a circulation pattern characterized by seawater flowing landward along the bottom and seaward along the surface.

Reflecting such considerations, the degree of mixing in an estuary is strongly influenced by the relative magnitude of water volumes relating to the river flow and the tide. This may be characterized by the ratio of the volume of freshwater entering the estuary over a tidal cycle to the water volume increase in the estuary from low tide to high tide. When this ratio is 0.1 or less, the tidal flow dominates a more modest river flow, resulting in a well-mixed estuary; when this ratio is about 0.2–0.5, the tidal flow is insufficient to produce mixing to the same extent, resulting in a partially stratified estuary; finally, when the ratio is greater than about 0.5, the tidal flow is modest relative to the river flow, so that the freshwater and saltwater tend to remain separate, resulting in a highly stratified estuary.

10.7 Estuarine Flushing

An important aspect of mixing in estuaries relates to the extent to which waters within an estuary are renewed, a process referred to as estuarine flushing. More rapid flushing promotes the ecological health of an estuary by causing improvements with respect to water quality, dissolved nutrients, dissolved oxygen depletion levels, suspended sediments, and algae blooms. The mechanisms leading to estuarine flushing relate to mixing resulting from tidal flows over each tidal cycle, referred to as tidal flushing, and to mixing resulting from freshwater from a river discharging into an estuary.

10.7.1 Flushing Parameters

Flushing time, which is a key measure of flushing, and parameters that influence tidal flushing are outlined below.

Flushing time. Flushing time is commonly used as a measure of the extent of flushing. This is the time taken to replace the freshwater in an estuary at a rate equal to the river flow rate; or, equivalently, the average time for freshwater entering the estuary to be discharged to the ocean. The flushing time T_F can be expressed as $T_F = V_f/Q_f$, where V_f is the volume of freshwater within the estuary at high tide and Q_f is the freshwater inflow rate. V_f as a fraction of the estuary volume V (at high tide) is not known *a priori* but may be estimated from information on the salinity distribution within the estuary. In particular, this ratio V_f/V may be approximated as

$$\frac{V_f}{V} = \left(1 - \frac{\overline{S}}{S_o}\right)$$

where S_o is the salinity of the ocean water and $\overline{S}$ is the mean salinity within the estuary. Applying this to the definition of flushing time, T_F may be expressed as

$$T_F = \frac{V}{Q_f}\left(1 - \frac{\overline{S}}{S_o}\right)$$

The application of the above equation depends on measurements of salinity within the estuary or on suitable assumptions relating to salinity.

Depending on particular circumstances, alternative definitions of flushing time have been adopted; also, a related measure referred to as *residence time* is sometimes used.

Tidal prism. A widely used parameter for tidal flushing is referred to as the tidal prism, denoted Δ. This is the increase in volume of water within an estuary from low tide to high tide. The larger the tidal prism relative to the overall volume of water in the estuary, the greater the opportunity for tidal flushing to occur. The following mechanism causes tidal flushing. A tidal flow causes "new" ocean water to be brought into an estuary during each flood tide, depending on the magnitude of the tidal prism. This mixes with and dilutes pollutants within the estuary, with some of the diluted pollutant then removed during the subsequent ebb tide.

Tidal exchange ratio. However, the "new" water entering the estuary during a flood tide is usually not entirely new in that it will include some estuarine water that had left the estuary during the previous ebb tide. This characteristic is described by the tidal exchange ratio R, which is the ratio of the volume of "new" ocean water relative to the total volume of water entering the estuary from the ocean during a flood tide. R depends strongly on a prevailing longshore current that removes any water discharged during an earlier ebb tide. If R is close to 1, most of the water leaving the estuary is removed, resulting in an improved level of flushing. Conversely, if R is small, then flushing occurs to a lesser extent, since removed water is largely returned to the estuary. For a given location, an estimate of R may be developed in terms of different water volumes and salinities. Specifically, R may be estimated from the following formula:

$$R = \left(\frac{Q_f T}{V_s}\right)\left(\frac{S_e}{S_o - S_e}\right)$$

Here, V_s is the volume of ocean saltwater entering the estuary during a flood tide, $Q_f T$ is the volume of freshwater discharged by the river in a tidal cycle, where T is the tidal period (i.e. including both a flood tide and an ebb tide), S_o is the salinity of the ocean water, and S_e is the average salinity of the water leaving the estuary during an ebb tide. The application of the above equation depends on suitable measurements or assumptions that enable an estimate of S_e to be made.

10.7.2 Selected Cases of Flushing

Taking account of the above, estuarine flushing may be assessed first by the flushing time and then, if possible, by the spatial and time variations of pollutant concentrations. This assessment may be sought for differing situations, ranging from a small harbor with no river sources and an instantaneous, single discharge of pollutant to a large estuary with complex characteristics, multiple river sources, and one or more sources of pollutant entering the estuary that are instantaneous and/or continual. Approaches to investigating a few such cases are now indicated.

Concentration reduction due to tidal flushing. Perhaps the most fundamental example relates to a quantity of pollutant being introduced into a harbor that has no freshwater inflow. (It should be noted that, while the decrease in pollutant concentration is of interest here, the concept of flushing time does not relate to this case of no freshwater inflow.) The relevant equation describing the decrease in concentration may be set up by assuming complete mixing within the harbor during each tidal cycle and considering the volume of "new" water entering the harbor during each flood tide, $R\Delta$, where, again, R is the tidal exchange ratio and Δ is the tidal prism. Such an analysis leads to a decrease in the average concentration in each tidal cycle, with the ratio of concentrations in successive tidal cycles expressed as

$$\frac{c_{n+1}}{c_n} = 1 - R\left(\frac{\Delta}{V}\right)$$

where V is the volume at high tide. This analysis may be extended to take account of a river inflow with a volume $Q_f T$ of freshwater each tidal cycle. In such a case, the equation for the concentration ratio is modified to

$$\frac{c_{n+1}}{c_n} = 1 - R\left(\frac{\Delta}{V}\right) - \left(\frac{Q_f T}{V}\right)$$

R may be reduced to anticipate incomplete mixing of incoming ocean water with harbor water.

Flushing time – river discharge predominant. Another fundamental example provides a simple estimate of the flushing time for an estuary with a river inflow, when tidal flushing plays a minor role. Assuming complete mixing across each section of a long, narrow estuary the salinity can be taken to depend only on distance from the estuary mouth. In one such approximation, described by Fischer et al. (1979), this mixing is associated with the longitudinal dispersion coefficient D_L and leads to an approximation in which the salinity decays exponentially with distance from the mouth of the estuary. This approach leads to the following expression for the mean salinity to ocean salinity ratio:

$$\frac{\overline{S}}{S_o} = \frac{[1 - \exp(-Q_f\ell/AD_L)]}{Q_f\ell/AD_L}$$

where ℓ is the length of the estuary and A is the cross-sectional area. This result can then be applied to the formula for the flushing time given above.

Flushing time – tidal discharge predominant. On the other hand, when tidal flushing is predominant, a different approximation may be adopted based on the tidal prism concept, now assuming complete mixing of incoming river water and of ocean water with estuarine water during a flood tide. This approach is reliant on the tidal exchange ratio R, such that some of the incoming ocean water is new and some is returning estuarine water. In its simplest form, this leads to the following expression for the flushing time T_F:

$$T_F = \frac{V}{(1 - R)Q_f + 2R\Delta/T}$$

(Alternate expressions have been developed, depending on the specifics of the assumptions made.) This estimate represents a lower bound because of the assumption of complete mixing.

One-dimensional models. For a long, narrow estuary, it is possible to extend the tidal prism concept to a series of segments along the estuary, with suitable relationships established between one segment and the next in relation to flows, pollutant transfers, and concentration levels. That is, the flows, salinity, and concentrations are averaged across each section, so that they are assumed to depend only on distance from the river mouth. Based on suitable assumptions regarding longitudinal mixing, a one-dimensional analysis can be undertaken to obtain estimates of the flushing time and the longitudinal variation of concentration.

Three-dimensional models. Eventually, a more comprehensive three-dimensional model may be developed and used to estimate flushing times and also to develop spatial and temporal variations of concentrations, taking account of three-dimensional bathymetry, different degrees of stratification and mixing, and varying assumptions regarding river flows and pollutant sources.

10.8 Salinity Intrusion in Estuaries

As a river brings freshwater to an estuary, the saltwater from the ocean intrudes beneath the freshwater and can extend some distance upriver. Thus, one characteristic of stratified flows in an estuary relates to the extent of saltwater intrusion into a river and potential changes to this pattern. This topic is important since changes to salinity intrusion patterns may lead to the loss of freshwater vegetation and habitat and may impact freshwater aquifers that are used for drinking water. Changes to the intrusion pattern may be brought about by interventions to the river or estuary, for instance, to improve navigation, develop flood control infrastructure, or introduce sedimentation control measures. Furthermore, sea level rise due to climate change is also expected to modify salinity intrusion patterns.

The most fundamental case arises with little or no tide and a highly stratified estuary. This leads to salinity intrusion in the form of a wedge-shaped bottom layer, containing denser saltwater, that hardly mixes with the overlying freshwater layer and is referred to as an *arrested saline wedge* or *arrested salt wedge*. An estuary characterized in this way may be referred to as a salt-wedge estuary. While some mixing occurs at the boundary between the two water masses, this is generally modest.

The intrusion length along the river bottom (as measured from the river mouth) and the wedge profile are of primary interest and may be estimated. Such a flow, which is influenced by a density difference and for which both momentum and buoyancy are important, is characterized by the densimetric Froude number, denoted F_Δ. This is defined as $F_\Delta = U/\sqrt{g'd}$, where U is the river velocity, d is the water depth, and

g' is an effective gravitational constant associated with the density difference, referred to as the *reduced gravity*. This is given by

$$g' = g\left(\frac{\rho - \rho_f}{\rho}\right)$$

where ρ is the density of the saltwater and ρ_f is the density of freshwater.

A simplified analysis may be undertaken by assuming a two-dimensional flow and developing continuity and momentum balance equations involving hydrostatic pressure differences and interfacial shear stresses. Factors that may be taken into account include friction on the seabed and entrainment across the interface.

For modest tidal flows, the wedge moves upriver or downriver over a tidal cycle. However, with an increasing tidal range, there is a greater degree of mixing and the distinctiveness of the wedge becomes weaker. The more general case can be treated by numerical modeling, although this is dependent on suitable assumptions relating to mixing-related parameters.

10.9 Turbulent Jets and Plumes

Turbulent jets and plumes have coastal engineering applications with respect to diffusers and marine outfalls. For such flows, the velocity and pollutant concentration fields associated with the discharge of an effluent are of interest and need to take account of advection and diffusion of the effluent within the ambient fluid, momentum and buoyancy effects, turbulent mixing, and turbulent entrainment (i.e. the transport of ambient fluid across the fluid interface). In this context, an effluent refers to the liquid that is discharged from an outfall, as distinct from the pollutant that is contained within the effluent. A jet refers to a discharge dominated by its initial fluid momentum, while a plume refers to a discharge driven by the buoyancy of the effluent relative to the surrounding fluid.

10.9.1 Jet and Plume Behavior

The behavior of jets and plumes is dependent on environmental parameters relating to the receiving water, the geometry of the source, and jet or plume parameters. Environmental parameters relate to the nature of the stratification, the ambient current field, and the turbulence characteristics of the flow. Geometric parameters relate to the shape, orientation, and location of the nozzle, port, or orifice. Primary jet/plume parameters, which refer to the conditions at emission, include

- specific mass flux: $Q = UA$ (units of m^3/s)
- specific momentum flux: $M = U^2A$ (units of m^4/s^2)
- specific buoyancy flux: $B = Qg'_o$ (units of m^4/s^3)

where U is the velocity of effluent at emission (assumed uniform across the jet/plume), A is the cross-sectional area of the nozzle, Q is the volume flow rate of effluent, g'_o is the reduced gravity at emission, given as $g'_o = g(\rho - \rho_o)/\rho$, ρ_o is the density of effluent at emission, and ρ is the density of the ambient fluid.

The following terminology is used to categorize jets and plumes:

- simple jet: driven by momentum flux only
- simple plume: driven by buoyancy flux only
- buoyant jet: both momentum and buoyancy flux are important, although buoyancy effects eventually dominate

The behavior of jets and plumes has been widely studied. In describing such flows, there is a distinction between the near-field that is typically considered to extend up to about 6–10 diameters away from the nozzle, a transition area, and the far-field some distance away. The following describes the far-field for a simple jet or plume discharging effluent steadily from a nozzle in the absence of a prevailing current.

Figure 10.4 provides a definition sketch relevant to the far-field, with z denoting distance from the nozzle and r denoting radial distance from the axis of the jet or plume. Since the jet or plume is axisymmetric and the flow is steady, the axial flow velocity w and the pollutant concentration c are taken to vary with z and r only. The figure gives an indication of how the axial velocity w varies with r for a given z and how its maximum value w_m reduces with increasing z. The figure also indicates how fluid entrainment contributes to a reduction of concentration with increasing distance from the nozzle.

The variation of $w(r, z)$ in the radial direction has been approximated as Gaussian and so may be expressed in the form:

$$w(r, z) = w_m(z) \exp\left[-\left(\frac{r}{b_w(z)}\right)^2\right]$$

where w_m is the maximum value of w that occurs along the jet or plume axis $r = 0$ and b_w is the half-width of the jet or plume, defined as the radial distance at which the velocity has reduced to $0.368w_m$ (corresponding to $w/w_m = \exp(-1)$). The concentration $c(r, z)$ follows a corresponding distribution,

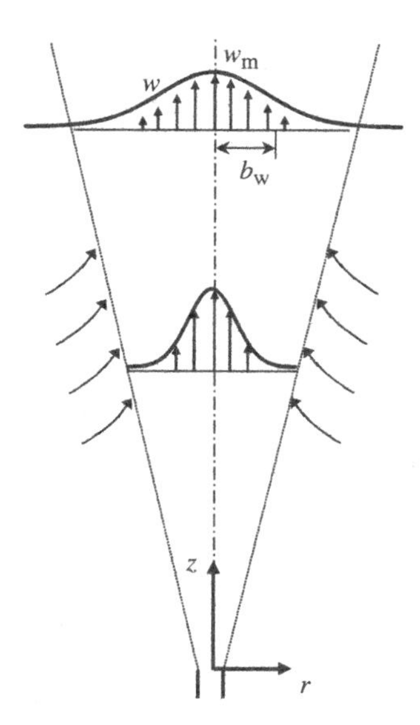

Figure 10.4 Definition sketch of a simple jet or plume.

with the half-width denoted b_c. In fact, the half-widths are proportional to z, corresponding to a linear increase in the jet or plume width with distance from the nozzle.

Available results for a simple jet and simple plume have been described by Fischer et al. (1979) and provide information on the maximum velocity $w_m(z)$, the maximum concentration $c_m(z)$ relative to the concentration at discharge, c_o, and the half-widths b_w and b_c.

For a simple jet:

$$w_m = 7.0\left(\frac{M^{1/2}}{z}\right)$$

$$\frac{c_m}{c_o} = 5.6\left(\frac{Q}{M^{1/2}z}\right)$$

$$b_w = 0.107\,z$$

$$b_c = 0.127\,z$$

For a simple plume:

$$w_m = 4.7\left(\frac{B^{1/3}}{z^{1/3}}\right)$$

$$\frac{c_m}{c_o} = 9.1\left(\frac{Q}{B^{1/3}z^{5/3}}\right)$$

$$b_w = 0.100\,z$$

$$b_c = 0.120\,z$$

The case of a buoyant jet is more complex, in part because both M and B are then relevant. The behavior of a buoyant jet at a distance z from the nozzle depends on z as well as on all three parameters Q, M, and B. Jet-like behavior is reliant on M and Q but not on B, while plume-like behavior is dependent on B and Q but not on M. Thus, the dominance of one form over the other depends on the distance z and the relative magnitudes of B and M. By dimensional reasoning, the dominance of one form over the other may be shown to depend on the dimensionless parameter $zB^{1/2}/M^{3/4}$. Thus, jet-like behavior occurs closer to the nozzle as momentum effects are more dominant, specifically when $z \ll M^{3/4}/B^{1/2}$, while plume-like behavior occurs further away when $z \gg M^{3/4}/B^{1/2}$. That is, sufficiently far from the nozzle, plume-like behavior eventually dominates.

10.9.2 Diffusers

A summary description of marine outfalls and diffusers is now provided. Typically, effluent is emitted through a series of ports or diffusers near the end of a submarine pipeline or outfall, as depicted in Figure 10.5.

The diffuser section is located so as to ensure that effluent is not transported back to the local shoreline. Thus, an outfall typically extends a few kilometers from the shoreline to a depth of 50 m or greater. The outfall itself may be buried beneath the seabed or lie on the seabed. The diffuser section typically relies on multiport diffusers, with ports designed to assure rapid dilution and dispersion of the effluent. Alternative designs are possible with respect to, for example, pipe diameter, the length of the diffuser section, the number and spacing of diffusers, and the specific port design. In some cases, port diameters or pipe sizes

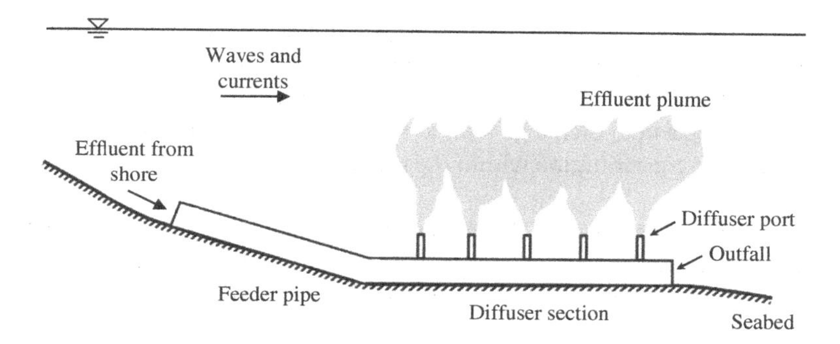

Figure 10.5 Schematic of an effluent plume associated with a multiport diffuser at the end of an outfall.

vary along the diffuser section so as to ensure an approximately uniform distribution of discharge among the diffuser ports.

As mentioned, the near-field flow in the vicinity of the individual ports is distinct from the far-field flow considered earlier. In the near-field, strong mixing and entrainment take place, associated with turbulence and relative buoyancy effects. A more detailed modeling of the near-field flow takes account of the jet discharge velocity from each port, the port diameter and spacing, the density difference between the effluent and the receiving water, and the flow direction and magnitude of the receiving water. The jets from the individual ports merge rapidly, so that, in the far-field, simplifications are made.

Problems

10.1 A mass m of pollutant is released instantaneously at location $x = x_1$ and time $t = t_1$ in a long narrow river that has a cross-sectional area A and an average current U. Develop an expression for the concentration $c(x, t)$ at a general location x and time t in terms of $M = m/A$, the diffusion coefficient D, U, x_1, and t_1. A mass of 6 kg is released in the river at location $x = 0$ m and time $t = 0$ min and a further mass of 4 kg is released at location $x = 300$ m and time $t = 6$ min. Use the principle of superposition to consider the concentration $c(x, t)$ in the river due to both discharges. If $A = 150\ \mathrm{m}^2$, $D = 1\ \mathrm{m}^2/\mathrm{s}$, and $U = 1$ m/s, plot the time variation (from $t = 10$ min to $t = 20$ min) of the concentration (in $\mathrm{g/m}^3$) at a location $x = 800$ m due to the first discharge, the second discharge, and the combined discharges. Hence, determine the maximum concentration at $x = 800$ m and the time at which this occurs.

10.2 ■ A mass of pollutant is released instantaneously on the seabed at a location where there is steady, uniform current U. The water has a uniform depth and is sufficiently shallow for the relevant two-dimensional advection–diffusion equation to apply. The concentration variation at a location x downstream of the source is being monitored. The concentration after time $2t$ is recorded as α times the value after time t. Develop an expression for the x-ward diffusion coefficient D_x in terms of U, t, x, and α. Hence, estimate D_x for the case $U = 0.4$ m/s, $x = 200$ m, $t = 4$ min, and $\alpha = 0.6$.

10.3 A marina has an average surface area of 84 000 m^2, an average depth at low tide of 5 m, a tidal range of 2 m, and a tidal period of 12 hr. A creek flows into the marina with a constant discharge of 1 m^3/s. A mass of effluent is discharged into the marina and is presumed to mix fully with the water within the marina. Assuming a tidal exchange ratio $R = 0.4$, estimate the number of days it would take for the average effluent concentration within the marina to be halved.

10.4 ■ An estuary is approximated as having a length of 5 km, an average width of 500 m, an average depth at low tide of 10 m, a tidal range of 3 m, and a tidal period of 12 hr. The river flow rate is 8 m^3/s. Based on the tidal prism concept and assuming a tidal exchange ratio of 0.4, estimate the flushing time for the estuary.

10.5 An outfall discharges effluent vertically at a rate of 0.8 m^3/s through a circular nozzle with a diameter of 0.45 m. The pollutant concentration at the nozzle is 10 g/m^3. The effluent density is 1000 kg/m^3 and the ambient fluid density is 1025 kg/m^3. Estimate the maximum flow velocity w_m and the maximum pollutant concentration c_m at a distance of 50 m from the nozzle, assuming that the discharge is a simple jet and a simple plume in turn. Develop a plot comparing w_m as a function of distance z for both a jet and a plume. At approximately what distance from the nozzle does w_m for a plume exceed that of a jet?

11

Design of Coastal Infrastructure

In this chapter, consideration is given to the design of coastal structures, harbors, and marinas. A general outline of the design process is initially provided, approaches to accounting for uncertainty are indicated in Section 11.2, and selected tools that may be utilized in the design process are described in Section 11.3. While these topics relate to engineering projects more generally, they are outlined in the context of coastal engineering.

11.1 The Design Process

The design process entails the establishment of objectives, criteria, and constraints. The objectives establish the overarching purpose of the infrastructure (e.g. to protect a site from coastal flooding). A related functional analysis may be undertaken so as to identify the key functions and objectives that the infrastructure is intended to fulfil. These may include, for example, adequate protection of vessels from waves, erosion control requirements, or flood protection requirements. Design criteria refer to the specifications and requirements to which the design should adhere with respect to safety and performance. They are based on codes, standards, regulations, analyses, and experience. Design constraints refer to limitations, restrictions, or conditions that must be satisfied by a design. Together, the criteria and constraints may relate to legislation, regulation, permitting, codes, and standards, as well as to technical, economic, health, safety, environmental, and societal factors.

A design is usually undertaken in different stages. These include a *conceptual design* or a feasibility study, sometimes referred to as a schematic design; a *preliminary design* with a focus on a framework that allows for flexibility, so that design changes may be made with relatively minor cost implications; and a *detailed design* that provides complete descriptions of all component parts of the project, through detailed drawings and specifications sufficient to proceed to construction and completion.

For larger projects, engineering design stages may be referred to as *Concept Select* or *Concept Design*, *pre-FEED* (Preliminary Front-End Engineering Design), and *FEED* (Front-End Engineering Design). The Concept Select stage evaluates different design options and selects the one that best meets the objectives and requirements. The pre-FEED stage refers to a preliminary stage in which the project's concept is developed, a detailed project scope is outlined, and a roadmap for the entire lifecycle of the project is formulated. The pre-FEED stage may refer to the *Basis of Design* that documents the principles, assumptions,

An Introduction to Coastal Engineering, First Edition. Michael Isaacson.

Companion website: www.wiley.com/go/coastalengineering

rationale, criteria, and considerations used for calculations and decisions required during design, and/or a *Design Statement* that provides the overarching objective or goal, standards and regulatory requirements, health and safety considerations, and technical, economic, environmental, cultural, and societal issues. The FEED stage completes the detailed design through detailed drawings and specifications sufficient to proceed to construction and project completion.

11.2 Accounting for Uncertainty

11.2.1 Kinds of Uncertainty

A key feature of virtually any design project in coastal engineering relates to identifying and understanding potential uncertainties and making allowances for these in the design. Uncertainties may be associated with the short-term and long-term randomness of variables associated with environmental loads (e.g. waves, wind, and water levels), the potential occurrence of rare events of uncertain magnitude (e.g. vessel impact, tsunamis, and earthquakes), and uncertainties in future trends that may not be revealed by an analysis of past conditions (e.g. sea level rise). However, while structural and material parameters (e.g. material properties, dimensions, connection details, reinforcing, etc.) and sediment and soil characterizations may be uncertain, these are usually represented by assumed parameter values or ranges. The general approach to accounting for the various kinds of uncertainties is outlined below, while Section 11.3.1 provides a refined approach based on an analysis of the probability of failure.

Finally, uncertainties may also arise with respect to the potential occurrence of detrimental events or hazards, which may relate to the kinds of variables indicated above or to a broader range of circumstances with respect to, for example, societal, environmental, economic, regulatory, or political considerations. Section 11.3.2 describes a risk analysis and risk management process relating to the potential occurrence of detrimental events or hazards.

11.2.2 General Approach

In recognizing the variability in the multiple parameters influencing a design, the most common design approach entails two component steps. The first seeks out various worst-case scenarios or design events, commonly selected on the basis of a specified return period. It then establishes criteria relating to each of these scenarios, usually associated with different modes of failure and/or different combinations of loads and analyses each of these in turn.

A second step incorporates adequate margins of safety, typically supported by relevant standards and guidelines. A load factor approach is commonly adopted in structural and geotechnical engineering design, as identified, for example, in ASCE (2022). This may take account of several kinds of loads acting simultaneously and is described further in Section 11.3.1. In a related way, the inclusion of adequate freeboard provides a margin of safety when a potential failure corresponds to the extent of coastal flooding.

As an extension to the general approach described above, uncertainties in future trends that may not be revealed by an analysis of past conditions, usually in the context of climate change, are accounted for by estimating conditions at a specified end-year that usually corresponds with the end of the project's design life.

As an indication of the above approach with respect to structural loading, the significant wave height with a specified return period is used to obtain the single largest wave condition, the resulting loads are obtained, and a load factor is then applied.

11.2.3 Extensions to the Approach

While the kind of approach outlined above is well established, there are two extensions that may be incorporated to assure that all scenarios are considered in a consistent manner.

- ***Encounter probability***. First, the encounter probabilities of different design events may be estimated in the context of comparing alternative scenarios. The encounter probability describes the probability of the design condition being exceeded during the design life of the project and is described in Section 4.5.5. Such considerations may guide the selection of return period. While typical encounter probability percentages may appear to be unduly large, it is emphasized that they refer to the design event and not to the load that is eventually used in a design.
- ***Probability of failure***. The calculation procedures, load factors, and load combination factors outlined in design standards are usually sufficient to undertaking a design. However, when circumstances are not suitably covered by a design standard, or when additional insights are needed, then a reliability analysis is undertaken in order to estimate the probability of failure, which may in turn lead to design adjustments. Such an approach is outlined in Section 11.3.1.

11.3 Selected Design Tools

Prior to considering aspects of the design of coastal infrastructure, an overview of selected design tools is provided. The following are considered in turn: a reliability analysis that is used to determine the probability of failure; a risk assessment and risk management procedure that analyses risks and considers approaches to mitigating risks; permitting and approval processes; a formal decision-making approach; and optimization modeling. While these are not coastal engineering topics, they are outlined here in a general way since one or several of these may be relied upon in undertaking a coastal engineering project. It is emphasized, however, that a design may not involve many or most of these, especially for smaller projects.

11.3.1 Probability of Failure

In some circumstances, a design may be enhanced by an assessment of the probability of failure. A reliability analysis that leads to such an assessment is now outlined.

In the context of structural engineering, *reliability* corresponds to the probability that the project will perform its required function under given conditions and is given as one minus the probability of failure under a combination of random or uncertain design variables. A *reliability analysis* determines this probability for different modes of failure or nonperformance scenarios, each depending on a set of input random variables whose probabilistic properties are known. In a sense, a reliability analysis corresponds to determining the probability of a consequence or an "output", i.e. the onset of failure, given the probabilistic properties of multiple random inputs.

Initially, however, the concept of failure in a coastal engineering context is considered. In structural engineering, "failure" usually refers to a structural or geotechnical failure and is distinguished from nonconformance whereby the structure is performing inadequately, for example, because of excessive displacements. However, in the present context, a failure is understood to include the latter kind of shortcoming, for example, with respect to excessive overtopping of a dike crest, or to excessive wave transmission past a floating breakwater.

A general outline of potential modes of failure for different kinds of coastal structure is provided in Section 11.4.1. Some modes of failure are progressive in that they may occur to different degrees of severity. For example, excessive overtopping may be considered to be a form of progressive failure.

A summary of approaches to undertaking a reliability analysis is now indicated. As mentioned, a reliability analysis determines the probability of a particular mode of failure occurring within the design life, dependent on a set of input random variables whose probabilistic properties are known or estimated. In a structural engineering context, this is undertaken by considering the capacity or "strength" of a structure, denoted C, needed to accommodate an applied load or "demand," denoted here as X, both of these relating to a particular mode of failure. A performance function G (also termed a failure function or reliability function) is then defined as $G = C - X$, such that the probability of failure corresponds to the probability that $G < 0$. Or, equivalently, the reliability corresponds to the probability that $G > 0$. Each of C and X depend on a series of variables with known probabilistic properties, thus enabling the probability of failure corresponding to $G < 0$ to be determined. This concept can readily be extended beyond a loading condition. Thus, with respect to the dike overtopping example, the capacity C to accommodate flooding corresponds to the dike crest elevation h, while the applied load X corresponds to the flood elevation that occurs. Thus, the probability of failure corresponding to $G < 0$ is given by the probability that $X > h$.

For situations that are suitably covered by codes and standards, in particular through prescribed procedures, load factors, and load combination factors, a designer may rely on such codes, which obviates the need to undertake a reliability analysis. For situations that are not suitably covered by available codes, a reliability analysis may be undertaken in various ways. In some cases, the probability of failure may be developed from a more fundamental analysis based on a defined failure criterion and on known or assumed relationships between the relevant variables. As an example of such an approach, a calculation of the probability of overtopping has been indicated in Section 7.9.5. More commonly, established numerical methods may be used, such as the First Order Reliability Method that determines a "most probable point" combination of random variables that influence failure. Another approach is based on a Monte Carlo simulation, which imitates the randomness of the output parameter (in this case, the occurrence of failure) through a large number of simulations, with each sample based on a value of each input random variable drawn from its known probability distribution. The resulting values of the output variable are then used to ascertain the probability of failure.

In coastal engineering, a reliability analysis may be undertaken to support a design in various ways. Examples include determining appropriate combinations of different simultaneous loads (e.g. due to waves, ice, and vessel impacts) that may contribute to a failure; estimating the probability of an instability failure of rock armor in a particular storm or over the design life; and developing a dike crest elevation based on prescribed probabilities of different extents of overtopping.

←■

11.3.2 Risk Assessment and Management

A *risk assessment* refers to the identification and quantification of the risks of a project, while *risk management* refers to approaches to monitor, eliminate, mitigate, and/or control risks. A risk assessment and risk management procedure may be undertaken for some coastal engineering projects, and a general outline of these is now provided.

Hazards. A hazard is an unwanted physical, societal, economic, or political event that could harm a project or its objectives. In general, hazards may include those relating to natural hazards (e.g. winds, earthquakes, and storms); environmental hazards (e.g. site conditions and environmental impacts); project management (e.g. scheduling delays, cost overruns, and scope creep); constructability; operational and technical issues (e.g. as relating to equipment and materials); and/or they may be reputational (e.g. loss of public or stakeholder confidence), regulatory (denial of approval to proceed), financial/economic/business (e.g. economic downturns, bankruptcy), technological (e.g. an unproven new technology), and political (e.g. changes in government taxation or other policies). Typically, only a subset of these may require consideration in a coastal engineering project.

Risk levels. In a risk assessment, a set of risks are quantified such that the risk level of each is defined as the likelihood of a hazard occurring multiplied by the magnitude of the consequence of that hazard occurring.

Likelihood and consequence levels. A common approach to developing likelihood and consequence levels, sometimes referred to as scores, relies on a five-point scale for each of these, based on the judgement of participants. These scales are indicated in Table 11.1.

Consequence categories. When assigning the consequence level associated with a hazard, it is possible than any one hazard may have different consequence categories, for example as relating to project costs, project schedule, project scope, health and safety, environmental impacts, and reputation. In determining risk levels, the assigned consequence level for any particular risk corresponds to that consequence category that leads to the most severe consequence level.

Refined rating scales. It is recognized that the five-point scales for likelihood and consequence are coarse approximations for calculating risk levels, since these scales assume an evenly distributed set of probabilities defining likelihood levels and an evenly distributed set of magnitudes defining consequence

Table 11.1 Five-point scales used for likelihood and consequence levels.

Assigned level	Description of likelihood	Description of consequence
1	Rare	Insignificant
2	Unlikely	Minor
3	Moderate	Moderate
4	Likely	Major
5	Almost certain	Catastrophic

levels. When hazard probabilities depart significantly from such an approximation (e.g. with respect to a tsunami that has a very low probability of occurrence), the five-point scales may be suitably modified or replaced by probability estimates. Likewise, when consequence magnitudes depart significantly from this five-point approximation, the scale may be suitably modified, most readily by estimating and using the effective cost of recovery or remediation if the hazard was to occur.

Risk classes. Individual risks may be categorized into different risk classes, such that they may be compared within a risk class but usually not across risk classes. Cost-related risks (more commonly referred to in the industry as "financial risks") are those risks for which the various consequence categories lead to additional costs to the project. Other classes of risk include "HSE risks" (health, safety and environmental risks) and regulatory and reputational risks.

Risk register. A risk register is a tabulation of the various risks that have been identified. It shows, for each risk, the likelihood, consequence, and risk level, as well as modifications to these that may arise as the various risks are mitigated or as circumstances change.

Risk matrix. A risk matrix depicts the relative severity of the various risks in the manner indicated in Figure 11.1. Thus, a risk matrix contains rows that correspond to likelihood levels, with values extending from 1 (lowest) to 5 (highest), columns that correspond to consequence levels, with values extending from 1 (leftmost) to 5 (rightmost), and cell entries that identify those risks corresponding to the various likelihood/consequence combinations. Thus, the most severe risks appear on the upper right of the matrix and the least severe risks appear on the lower left. This matrix represents a convenient visual for identifying these risks.

Mitigation measures. Beyond an initial risk analysis, there is often a need to consider potential mitigation measures for the most severe risks, leading to updated likelihood and consequence levels, and so to

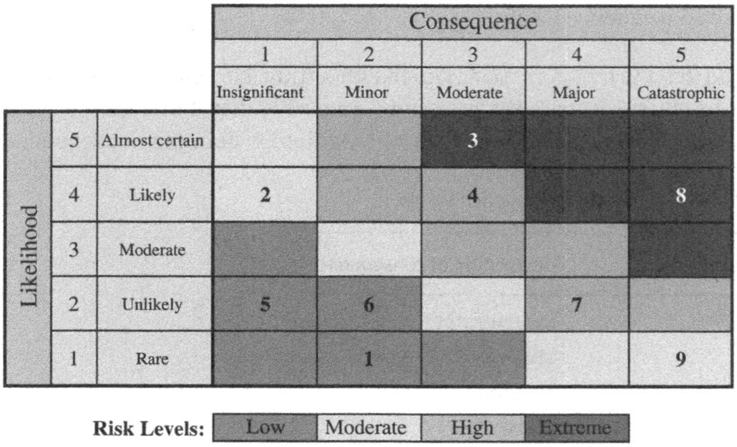

			Consequence				
			1	2	3	4	5
			Insignificant	Minor	Moderate	Major	Catastrophic
Likelihood	5	Almost certain			3		
	4	Likely	2		4		8
	3	Moderate					
	2	Unlikely	5	6		7	
	1	Rare		1			9

Risk Levels: Low | Moderate | High | Extreme

Figure 11.1 Illustration of a risk matrix.

updated risk levels. In a project's development, the risk register is updated periodically, with responsibilities assigned, as various mitigation measures are implemented and tracked.

11.3.3 Permits and Approvals

One aspect of the design process is the need to secure relevant permits and approvals for a project, and to modify designs accordingly. Different projects may require different permits or approvals issued by relevant authorities to enable a project to proceed, sometimes multiple permits are required and sometimes approvals from different agencies or levels of government may be necessary. These may relate to different federal, state, and municipal agencies and laws, including zoning bylaws, habitat protection requirements, environmental impact assessments, and a municipality's Official Community Plan. Approvals may require a review of the preliminary or detailed design in order to assure adherence to relevant codes, standards and guidelines, an assessment of societal and environmental impacts, and a required consultation process with relevant stakeholders.

Consultations. A design may be required to incorporate a mandated consultation process, which generally includes the preparation of an approval application and impact assessment, a distribution of these for stakeholder review and follow-up, discussions between the proponent and stakeholders, and subsequent decision-making and dissemination. This process may be undertaken through the development and execution of a consultation plan based on a specified set of requirements Typically, a consultation plan includes the following elements:

- ***Stakeholders***. An identification of stakeholders, including their roles and responsibilities.
- ***Formats***. An identification of communication formats with stakeholders, including noninteractive formats (e.g. websites, storyboards, and public notices) or interactive formats (e.g. surveys, public meetings, and charettes) and an identification of the number, timing, format, and groupings of the various consultations that are to occur.
- ***Utilization***. An articulation of approaches to responding to feedback, including discussions of feedback, the development of responses, and descriptions of how feedback may be utilized. For example, feedback may be dismissed with a suitable rationale, or may be incorporated into the process of selecting from several design options or of modifying the design.
- ***Communicate outcomes***. Approaches to a dissemination of the outcomes of the consultations.

Depending on the jurisdiction and context, there may be additional legal requirements for consultation with Indigenous peoples, distinct from stakeholders and affected communities.

Engagement. Many larger engineering projects entail an engagement process with communities, Indigenous peoples, and stakeholders as may be relevant. This is distinct from any consultation process that may be required. This is a process by which the project sponsors build ongoing, long-term relationships for the purpose of developing a collaborative culture of shared ownership and applying a collective vision for the benefit of the community and the sponsors. Community benefits may take various forms including a community's involvement in decision-making, employment or participation opportunities, and the development of social projects and support.

11.3.4 Decision-Making and Option Selection

A design may involve the selection of alternative designs or options, given that the selection depends on multiple criteria. For larger projects, this arises at the concept select stage. The decision-matrix method is a common approach to undertaking multi-criteria decision-making. In summary, the procedure entails the following steps: a set of criteria that impact the assessment of options is specified; weights are assigned to each criterion so as to express its relative importance, with weights usually selected so as to sum to 10; for each option, scores are assigned with respect to each criterion, bearing in mind that these may include both quantitative and qualitative scores; scores are normalized based on the lowest and highest possible scores for each criterion; and, finally, a weighted sum of all the scores is obtained for each option, leading to the selection of the option with the highest score.

This process depends on a suitable engagement of participants in establishing the weights for the different criteria, which may reflect differing priorities, and in developing scores for qualitative criteria, including those obtained through surveys and other forms of interaction with user communities and other stakeholders.

Extensions and alternatives to this method are possible. One such extension is referred to as the Kepner-Tregoe matrix method, also known as the KT-matrix method. This involves a series of steps that include an accounting of risks and a consideration of potential mitigation measures in addressing those risks.

11.3.5 Optimization Models

In some cases, it may be appropriate to develop an optimization model to support a design. An optimization model is used to determine an optimum set of design variables that meet one or more specified objectives, subject to one or more constraints. More specifically, an optimization model seeks to minimize or maximize one or more "objective functions," each of which corresponds to a mathematical representation of an objective to be achieved. These may relate, for example, to certain dimensions, weights, stresses, and effective costs. With respect to costs in particular, the model may, in effect, incorporate and extend a cost–benefit analysis.

The constraints indicate limitations that need to be adhered to while seeking to meet the objectives. They relate, for example, to allowable stresses, deflection limits, and costs. In order to solve an optimization problem, the relevant design variables, objective function(s), and constraint(s) are identified; then an appropriate algorithm is used to determine the best choice amongst a continuum of options.

In coastal engineering, optimization models may be applied to determine, for example, crest elevations and other dimensions of coastal defense structures, the scheduling and extent of sediment delivery with respect to a beach replenishment program, the optimum lengths and locations of groins and offshore breakwaters, and the optimum sectional design of a shoreline structure.

One such example corresponds to an optimization of the length and/or crest elevation increase of a dike upgrade. Either of these may be optimized by developing a suitable objective function and thereby minimizing the net present worth of the upgrade as a function of dike length and/or crest elevation. The net present worth would normally take account of direct costs (construction and increased annual maintenance costs), benefits (viewed as negative costs, e.g. lower insurance rates, increased property values), and the costs of potential flooding (probability of occurrence for different flooding scenarios multiplied by the estimated costs of damage for each scenario). As an extension to this example, the timing and extents of successive dike upgrades over the long term, taking account of sea level rise, can be optimized.

11.4 Aspects of the Design of Coastal Structures

This section outlines modes of failure for various kinds of coastal structure, used to guide design, summarizes a selection of design criteria that may be relevant, and then provides commentary on design practices with respect to coastal structures, including aspects of detailed design.

11.4.1 Modes of Failure

It is prudent to identify potential modes of failure of a coastal structure, since these may be used to establish corresponding design criteria and to verify that a design has suitably taken these criteria into account. Potential modes of failure may be identified on the basis of experience and knowledge gained from related projects.

For seawalls, caissons, and wall-type structures, modes of failure are associated with structural stability with respect to overturning, sliding, or settlement; structural integrity with respect to fracture, deformation, or fatigue; structural materials, including corrosion resistance and durability, especially in the context of saltwater exposure and intermittent wetting; and foundation failures with respect to soil conditions, bearing capacity, liquefaction, and scour.

For rubble-mound structures, a failure may be associated with unit instability, failure of the toe protection, possibly triggered by scour, and the structural integrity of individual units, recognizing that any of these could trigger the mass movement or slumping of armor units, either with respect to the surface layer or other failure planes within the structure. A failure may also be associated with erosion or damage to the structure crest and/or the sheltered side of the structure.

For pile-supported structures, modes of failure may include foundation failures of piles, fatigue failures, and failures at connections to the superstructure.

For an earthen embankment, modes of failure include erosion of different portions of the embankment surface, and piping (internal erosion resulting in voids within the soil) due to seepage, with the possibility that either of these could lead to a washout or a breach.

For floating structures, modes of failure relate to listing, capsize, and sinking. For long floating structures, connections between adjacent units may fail. For a pile-restrained floating structure, the supporting piles may fail from fatigue, a foundation failure, or a failure at the connections to the super-structure. For moored structures, the mooring system may fail in various ways, including a failure at the connections to the structure or a berth. The anchors may also fail, possibly through dragging.

11.4.2 Design Criteria

Design criteria with respect to coastal structures include metocean design criteria, structural/ geotechnical design criteria, usually with respect to various modes of failure, functional or operational criteria regarding the effective performance of the structure, and criteria relating to environmental and other impacts. Regulatory requirements may or may not overlap with specific criteria.

Metocean design criteria. Metocean design criteria refer to the specifications and requirements for a project with respect to parameters relating to environmental conditions that may influence the project's design. ("Metocean" is a contraction that refers to the combined effects of meteorological and oceanographic conditions.) For larger projects, a distinct metocean study is typically undertaken to provide such information. This includes information on the wind climate, including descriptions of both annual and extreme wind conditions, extreme wave conditions including significant wave heights for various return periods and possibly wave directions, annual wave conditions in the form of a scatter diagram, water levels and water level components including tides, storm surge and relative sea level rise, and descriptions of the current field. Depending on the requirements of a project, the study may also include information on tsunamis, ice conditions, water temperature, and meteorological parameters (e.g. precipitation and air temperature). Finally, climate change impacts may be considered to form a distinct set of criteria, for example with respect to sea level rise, ice free conditions, and a specified end-year at which these and other criteria are developed. Metocean parameters and data may be developed from field measurements, agency and private sector data portals (e.g. tides, wind data and sea level rise), and numerical modeling and analyses.

Structural/geotechnical design criteria. Structural/geotechnical design criteria may be developed on the basis of on the potential modes of failure that have been identified. For seawalls, caissons, and wall-type structures, these relate to structural stability, structural integrity, structural materials, and foundation criteria. For rubble-mound structures, they relate to structural integrity, structural materials with respect to individual armor units, the instability of individual units, slope failures, failure of the toe, and damage to the crest and sheltered sides. For an earthen embankment, these include surface erosion criteria, geotechnical criteria, and stability with respect to seepage, possibly taking account of seepage control measures. For a floating structure, these relate to freeboard, floatation, hydrostatic and hydrodynamic stability, and excessive motions with respect to resonances. Additional criteria relate to the connections between adjacent units and to the mooring system and anchors.

Functional or operational criteria. These may relate, for example, to the extent of wave overtopping of a shoreline protection structure, erosion and sediment transport with respect to erosion control structures, the extent of wave transmission past a floating breakwater, and excessive motions of a floating structure. The latter may be associated with, for example, pedestrian safety and comfort with respect to a floating dock and downtime limitations with respect to operational activities (e.g. weather-dependent construction windows, and cargo discharge/loading operations). Such criteria may refer to different return periods: for example, shorter return periods with respect to wave transmission past a floating breakwater, medium-term return periods with respect to a beach nourishment project, and longer return periods with respect to coastal flooding.

Finally, some design criteria relate to specific kinds of infrastructure. For example, additional criteria for flood barriers with retractable flood gates relate to navigation requirements with respect to the flood gate dimensions and elevations, including channel widths and sill or bottom elevations, and the frequency of flood gate closures, recognizing that closures will become more frequent in later years on account of sea level rise.

Other criteria. Environmental impact and other criteria may relate to sediment transport behavior, fish passage through gaps, water circulation or flushing, habitat preservation, enhancement or compensation, and effluent discharge. Some of these criteria are related to regulatory requirements.

11.4.3 Design Loads and Load Factors

A key aspect of the preceding outline is the requirement to determine design loads on coastal structures. Where possible, it is prudent to rely on various design standards and guidelines. The standard ASCE (2022) is particularly relevant with respect to structural engineering aspects. This standard defines "V-zones" and "coastal A-zones," where coastal flooding rather than riverine flooding may occur and wave conditions are sufficiently severe. It describes the determination of "flood loads" that include hydrostatic and hydrodynamic loads, wave loads, breaking wave loads, and impact loads resulting from debris, ice, and other sources, and it requires the use of a load factor of 2.0 with respect to flood loads. The standard also specifies an approach to considering the combination of flood loads, using a load factor of 2.0, along with other loads that may act simultaneously, including dead load, live load, and wind, snow, and rain loads, using load combination factors that range from 0.3 to 1.2.

Load calculations may need to be adapted to different circumstances. For example, for a long, fixed structure subjected to oblique waves, the spatial and time variations of loads may need to be taken into account, since an assumption of maximum wave loads acting simultaneously on the entire structure would lead to a significant over-design.

11.4.4 Detailed Design

The foregoing provides a general framework for undertaking a coastal engineering design, based on the application of design guides and the tools and methods that have been outlined and as may be relevant. Attention is now given to detailed design practices with respect to coastal infrastructure, with a focus on rubble-mound structures. Again, a number of design guidelines and standards relating to the coastal engineering aspects of detailed design may be relied upon. These include the Shore Protection Manual (1984), the Coastal Engineering Manual (2002), the Rock Manual (2007), the Coastal Construction Manual (FEMA 2011), and the EurOtop Manual (2018). As well, there are many other standards and guidelines that provide more specific information on different kinds of detailed design. Examples of British standards include the Beach Management Manual (CIRIA 2010), CIRIA guidelines for groins (CIRIA 2020), and a BSI standard (BSI 1991) with respect to the design and construction of breakwaters. There are many related standards and guidelines emanating from Europe and other countries, for example an ISO standard (ISO 2007) for coastal structures with respect to waves and currents. Other standards and guidelines focus on particular kinds of infrastructure, including berths, fendering and mooring systems, dredging and land reclamation, structures such as jetties, dolphins, pipelines, and floating structures. Finally, the book by Reeve et al. (2018) provides a helpful description of a selection of detailed design requirements.

Beyond a reliance on available standards and guidelines, companies often rely on previous in-house detailed designs and on in-house technical practice or other guidelines with respect to detailed design, especially when required information in the open literature is insufficient. Examples with respect to rubble-mound structures relate to rock gradation and filter layer specifications, which may be influenced by local conditions and the availability of materials.

It should be borne in mind that there are different extents to which a detailed design is undertaken, depending on the scale of a project and the magnitude of the consequences of failure. For example, Figure 8.6 illustrates differences in the level of detailed design with respect to a simpler design with respect to riprap protection of a small seawall subjected to more modest conditions, versus the design of a large breakwater used to protect a port that would be exposed to much more severe conditions.

For a rubble-mound structure, aspects of detailed design include the selection of armor unit size on the basis of stability considerations, the breakwater slope, the breakwater crest elevation and width, the number of armor layers, the armor size distribution, material sizes for underlayers, filter layers and/or the core, toe protection, differences between the breakwater head and breakwater trunk, low-crest and submerged breakwaters, and crest structures that include crown walls. Brief comments on a selection of these are provided.

Rubble-mound slope and stability. For a given water depth and design wave condition, a steeper slope requires less material and a lower footprint but a larger unit mass and/or units with a larger stability coefficient. However, a steeper slope also increases the reflection coefficient, which may be a consideration for some situations. The slope is often selected to be 1/1.5, 1/2, or 1/3, with preference of a steeper option unless requirements indicate otherwise. Beyond the fundamental equations provided in Section 8.3, unit stability and the extent of damage are influenced by a wide range of parameters relating to incident wave conditions, the shoreline profile, and the characteristics of the units and the rubble-mound section. The Coastal Engineering Manual (2002) contains a detailed discussion of such matters.

Gradation. At the detailed design level, specifications of armor size gradation for the outer layer or for the structure overall are normally developed in terms of the median mass (or corresponding nominal diameter) of the armor units, with the associated gradation then specified with respect to corresponding minimum and maximum masses, or, in greater detail, through a series of mass values corresponding to different "percentage finer" ranges. The gradation for other armor layers may be described in a similar manner.

Filter layers. Likewise, at the detailed design level, when a single filter layer is used as a base for a single-layer rubble-mound structure, the filter layer is specified with respect to the layer thickness, the median mass (or corresponding nominal diameter) of the filter materials, which is dependent on the median size of the layer being supported, and the associated gradation. The latter is specified in terms of minimum and maximum masses, or, in greater detail, through a series of mass values corresponding to different "percentage finer" ranges. If relevant, the specifications of a filter layer may be extended to multiple filter layers and/or a core layer.

Unit integrity. The design of very large armor units requires attention to the strength, material properties, and shapes of individual units in order to avoid structural integrity failures of individual units, which would trigger a failure of the overall breakwater.

Toe protection. Close attention is given to the detailed design of toe protection, recognizing the potential mode of failure associated with the toe. A range of alternative designs may be adopted, with Figure 8.6 illustrating one such example. Variations include the use of an apron or berm placed horizontally over the seabed; the use of a geo-filter that prevents washout of finer sediments; an excavated trench so that the toe protection is embedded within the seabed, often at the expected scour depth since the toe is susceptible to erosion; a toe bund, which is a large unit cluster at the base of the armor layer; and different armor sizing, layers, dimensions, and slopes.

11.5 Design of Harbors and Marinas

Aspects of the design of harbors and marinas are now considered. A harbor, which may be natural or constructed, is a generic term that refers to a location along the coastline where marine vessels may be moored and protected from severe storms. A marina refers to a constructed harbor that provides moorage for pleasure craft and small vessels. The terms "fishing harbor" and "small craft harbor" are also used. The former relates to fishing vessels, while the latter relates to pleasure craft and/or fishing vessels. A port refers to a harbor where ships take on or discharge cargo and/or passengers. The focus of this section is on the design of marinas, although some items relating to ports are mentioned. For specific information on the design of marinas and small craft harbors, the reader is referred to Tobiasson and Kollmeyer (1991) and ASCE (2012).

11.5.1 Design Considerations

The range of considerations commonly taken into account in the design of a marina is outlined here. Attention may be given to the size, number, categories, and characteristics of vessels to be accommodated; the degree of wave protection to be provided to vessels; extreme water levels that impact wave overtopping and shoreline or facility flooding; the design of a breakwater layout as influenced by wave conditions approaching the marina and wave protection requirements; the sectional design of a breakwater, including its structural stability and integrity, foundation, and crest elevation; the design of docks, slips, and other facilities; water quality and flushing; sediment-related issues, including potential accretion, erosion, and scouring; and navigational issues, including the orientation and width of a marina's entrance or entrance channel. Also, if relevant, attention may also be given to: the design of mooring systems for larger vessels; seiche and seiche-induced loads and motions of larger vessels; impacts of potential ice cover and icing; and exposure to a potential tsunami. Other considerations such as those relating to economics, land transportation, and other land-based activities are considered outside the scope of coastal engineering.

Most of the above considerations have been mentioned elsewhere in this text, so that this section focusses on a few that have not otherwise been introduced. In particular, attention is given to the required level of protection to be provided to vessels in a marina or small craft harbor, navigational issues, the impacts of potential ice cover and icing, and certain aspects of port design.

11.5.2 Acceptable Wave Climate

Various published criteria are available for the wave climate considered acceptable within marinas and small craft harbors, also referred to as wave agitation criteria or harbor tranquility criteria. Fournier et al. (1992) provided a useful review of criteria that have been established worldwide, with a focus on fishing harbors. An early rule of thumb was that wave heights within a marina should not exceed about 0.3 m. However, this does not give attention to locations within a marina, the size and category of vessels, the definition of wave height, and the probability of occurrence. Two more comprehensive sets of criteria are provided below.

American Society of Civil Engineers. ASCE (2012) has published wave agitation criteria with respect to acceptable wave conditions in small craft harbors as provided in Table 11.2. (This stems from work reported by Isaacson and Mercer 1982.) The table distinguishes between requirements for 1-wk, 1-yr, and 50-yr wave conditions, between head seas and beam seas, between short period and long period waves, and between "moderate," "good," and "excellent" conditions.

Fisheries and Oceans Canada. DFO's "Small Craft Harbours" (SCH) program has published wave agitation criteria for three categories of small craft harbor (see Fournier et al. 1992), defined as follows:

- Class A harbor: > 800 vessel-meters
- Class B harbor: 300 – 800 vessel-meters
- Class C harbor: < 300 vessel-meters

In the above, "vessel-meters" refers to the daily average of the lengths of all vessels in a harbor over the most active 31 consecutive days in a year. That is, the class of harbor is related to the number of vessels that can be accommodated, the average size of the vessels moored in the harbor, and the extent of usage. The criteria provide for allowable maximum significant wave heights as described in Table 11.3. The table note corresponds to the threshold wave heights being exceeded no more than 15, 76, and 153 hours per year for Class A, B, and C harbors, respectively.

Table 11.2 ACSE wave agitation criteria.

		H_s (m) for "good" conditions		
T (s)	Heading	$T_R = 1$ wk	$T_R = 1$ yr	$T_R = 50$ yr
< 2	(Any)	0.3	0.3	—
> 2	Head	0.15	0.3	0.6
	Oblique	$0.15 - 0.07 \sin\theta$	$0.3 - 0.15 \sin\theta$	$0.6 - 0.37 \sin\theta$
	Beam	0.08	0.15	0.23

Notes to table:
1. For $T > 6$ s, the following additional criterion is in effect: $d_o = 0.5$ m, 0.6 m, and 1.2 m for $T_R = 1$ wk, 1 yr, and 50 yr, respectively, where d_o is the horizontal orbital diameter of wave motions.
2. For "excellent" and "moderate" conditions, the H_s values are to be multiplied by 0.75 and 1.25, respectively.
3. In the table, θ is the wave approach direction relative to a head sea (i.e. $\theta = 0°$ for a head sea).

Table 11.3 SCH wave agitation criteria.

	H_s (m)	
Location	**All recreational boats; fishing boats < 15 m**	**Fishing boats > 15 m**
Harbor entrance	1.00	1.00
Mooring basin	0.50	1.00
Berthing area	0.25	0.50

Note to table: Listed H_s values are to correspond to the following exceedance probabilities $Q(H_s)$:

$$Q(H_s) = \begin{cases} 0.17\% & \text{for a Class A harbor} \\ 0.87\% & \text{for a Class B harbor} \\ 1.74\% & \text{for a Class C harbor} \end{cases}$$

It is noted that this form of criterion requires information on the year-round wave climate, which is not related to wave heights corresponding to various return periods as may be obtained by an extreme value analysis. However, Section 4.7 describes approaches to developing annual wave conditions that are relevant to the above criteria.

11.5.3 Navigation

Navigational issues relating to marina design include the following:

- The width of the entrance and/or entrance channel needs to be sufficient for navigable access, dependent on the sizes of vessels and on whether one-way or two-way traffic occurs.
- The water depth at low tide should be sufficient to avoid vessel grounding taking account of the maximum draft of the vessels using the marina.
- Vessel speeds near the marina entrance and within the marina should be limited in order to reduce wakes and vessel-generated waves, so as to assure public safety and comfort.
- If possible, the marina entrance should be oriented so as to minimize navigation near the entrance under beam seas. This objective is usually met since breakwaters are most often oriented normal to the prevailing wave direction.

11.5.4 Ice Cover and Icing

For northern regions, ice cover limits the extent of the open-water season. Therefore, wave hindcasting and an extreme value analysis should relate to the relevant wave conditions during the open-water season only. As indicated in Section 7.8, climate change is expected to result in a long-term increase in the open-water season, extending into the spring and fall when more severe winds occur, resulting in increased design wave conditions. Therefore, the impact of climate change on the length of the open-water season needs to be assessed.

Distinct from ice cover, icing may influence loads on harbor structures and may cause armor units of breakwaters to be dislodged.

11.5.5 Ports

While the focus of this section has been on marinas, many of the considerations are relevant to the design of ports, but with some changes of emphasis. Per Bruun (1981) provides a foundational text with respect to port engineering. Since ports relate to ships and larger vessels, they are able to accommodate berthed vessels exposed to more severe wave conditions than those considered acceptable for marinas. Breakwaters are only required for ports exposed to large ocean swell and so will be composed of large armor units.

In lieu of specified wave conditions at a berthed vessel's location, there is usually a need to undertake a mooring analysis of berthed vessels, taking into account the vessel characteristics and the details of the mooring/berthing system. The analysis is used to assess the mooring system's suitability, including vessel motions, line tensions relative to breaking strengths, and loads on line connections to bollards and cleats. The analysis may need to take account of seiche and low-frequency oscillations that may cause resonances in the mooring system. Such analyses are usually undertaken using a sophisticated mooring analysis model.

Problems

In lieu of a selection of problems relating to this chapter, this section contains three assignments relating to design. These may be adapted by the instructor with respect to different locations or circumstances. Note that responses to assignments 1 and 2 may need to rely on information beyond the summary information provided in this text.

Written Assignment – Risk Assessment and Mitigation

11.1 Your company is tasked with designing a shore protection scheme along a 1-km stretch of shoreline for a waterfront residential development, intended to protect against flooding, support recreational use, and meet environmental impact requirements. Your team has developed a preliminary design based on rubble-mound protection fronting a seawall, incorporating nature-based elements. The client has requested that you undertake a risk analysis of the project, so as to incorporate selected mitigation measures in order to complete a detailed design.

Carry out a hypothetical risk assessment of the project, taking account of risks relating to permitting (e.g. environmental approvals not granted), safety issues during construction, construction management, rubble-mound instability and/or toe erosion leading to slumping, stakeholder and Indigenous peoples' engagement and consultations, and reputational or legal risks. Develop a risk register and a risk matrix.

Consider approaches to mitigating the most severe risks, and thereby develop a mitigated risk register and risk matrix. Develop a brief memorandum describing your assessment and summarizing the mitigation measures that need to be adopted in developing the detailed design and design process.

Written Assignment – Option Selection

11.2 Your company has developed two options for the conceptual design of a breakwater protecting a small marina. One is based on a pile-supported timber wall elevated above the seabed, and the

- ***Finite difference methods***. In these methods, derivatives in the governing equations are approximated by differences of variable values across grid cells. A large number of schemes are available.
- ***Finite element methods***. These seek approximate solutions to partial differential equations or integral equations, based on a discretization of the physical domain and on approaches to minimizing an error function relevant to the problem, as deriving from the calculus of variations.
- ***Method of characteristics***. This method reduces a partial differential equation to a family of ordinary differential equations along suitable surfaces or curves, so that the solution can be integrated from an initial set of specified parameters.

Key features in the development of a numerical model include the ease of data specification, the extent of discretization required, numerical accuracy, numerical stability, the robustness of a solution to minor variations of input data, approaches to calibration and/or validation, clarity in user manuals, and clarity in the display of results, including graphical output.

Numerical models range from, on the one hand, fairly simple models developed on a spreadsheet that solve a particular equation or set of equations in order to address a single issue to, on the other hand, sophisticated and comprehensive models that incorporate multiple topics, demand high computational power, and require specific training in order to be used.

12.3 Model Laws

Laboratory models are used to simulate a full-scale situation on the basis of smaller scale model tests in the laboratory. Figure 12.1 shows an example of a laboratory model study used to assess downtime for a ship loading terminal. Model conditions are established to represent suitably the full-scale situation, so that measurements drawn from the model can then be used to obtain corresponding values at full scale.

Figure 12.1 Photograph of a model study to assess downtime for a ship loading terminal. *Source:* Reproduced with permission of National Research Council Canada/Conseil national de recherches Canada.

A prerequisite to the use of a laboratory model is the development of model laws that provide relationships between input or measured model-scale parameters (subscript m) and those for the full-scale situation being simulated, referred to as the prototype (subscript p). These relationships are conveniently expressed in terms of *scale factors*, where the scale factor k_x of a variable x is defined as x_m/x_p. Scale factors may be obtained by equating relevant dimensionless groups between model and prototype obtained via a dimensional analysis, by invoking required similarities between model and prototype, or by relying on defining relationships or governing equations between the variables. These three approaches are summarized below.

12.3.1 Dimensional Analysis

A dependent variable (e.g. a force) is expressed as a function of a set of independent variables (e.g. depth, wave height, gravitational constant, and fluid viscosity). A dimensional analysis reduces this to a relationship between a set of dimensionless groups. This may be written in the form $\pi_x = f(\pi_1, \pi_2, \ldots)$, where the πs are dimensionless groups that are product combinations of the relevant variables. For example, a force coefficient is expressed as a function of Reynolds number and Froude number. Usually, π_x contains the dependent variable x being investigated (e.g. a force), while the remaining πs contain the independent variables only (e.g. water depth, wave period, and wave height). Then, if $\pi_1, \pi_2, \ldots$ are each made identical between model and prototype, it follows that $(\pi_x)_m = (\pi_x)_p$. Therefore, a measurement of x_m can be used to determine the corresponding value of x_p. In practice, not all the πs may be able to be held constant between model and prototype (e.g. Reynolds number), and special steps may be needed to take account of this disparity.

A fundamental example of the use of this approach to developing relationships between scale factors is as follows. If one of the relevant dimensionless groups is the Froude number, defined as $U/\sqrt{gL}$, where U is a characteristic velocity, g is the gravitational constant, and L is a characteristic length, then constancy of the Froude number can be used to express the scale factor for velocity, k_U, in terms of the length scale factor k_L. First, the Froude numbers of the model and prototype are equated:

$$\left(\frac{U}{\sqrt{gL}}\right)_m = \left(\frac{U}{\sqrt{gL}}\right)_p$$

Since g is the same in the model and prototype, this readily yields $k_U = \sqrt{k_L}$. Therefore, if a model is built with, say, $k_L = 1/10$, then velocities are related by $k_U = 1/\sqrt{10}$.

12.3.2 Similarity

Similarity between model and prototype may refer to geometric similarity (lengths between model and prototype are in the same ratio), kinematic similarity (velocities between model and prototype are also in the same ratio), and dynamic similarity (different kinds of forces between model and prototype are also in the same ratio). Requirements of dynamic similarity lead to constancy of dimensionless groups between model and prototype, so that similarity requirements provide a possible alternative to a dimensional analysis. Using the same example as above, dynamic similarity involving gravitational and inertial forces implies constancy of Froude number. This is referred to as Froude similarity or similitude. As indicated above, this may be used to develop the scale factor for U in terms of the scale factor for L as $k_U = \sqrt{k_L}$.

12.3.3 Defining Relationships and Governing Equations

In a related manner, scale factors for some variables may be obtained by relying on defining relationships or governing equations between the relevant variables. As an example, since acceleration a is, by definition, a length divided by a time squared, the scale factor for acceleration, k_a, may be expressed in terms of the scale factors for length, k_L, and for time, k_T, as $k_a = k_L/k_T^2$. Likewise, fluid pressure p takes the form ρgL, where ρ is fluid density, g is the gravitational constant, and L is a length. If water with the same density is used in the model, then, given that $k_\rho = k_g = 1$, it follows that $k_p = k_L$.

12.3.4 Scale Effects

In general, it is not possible to achieve complete similarity between model and prototype because some quantities cannot be suitably scaled. This inability to simulate a particular variable or characteristic in a model with respect to scaling requirements is referred to as a scale effect. This situation may arise in various ways in coastal engineering modeling, for example, with respect to surface tension effects, frictional effects, sediment transport mechanisms, stratified flows, and so on. An elaboration of the common case relating to Reynolds number is given below.

12.3.5 Reynolds Number Disparity

Perhaps the most well-known example of a scale effect arises from an attempt to suitably scale gravitational, inertial, and viscous forces simultaneously. This imposes a requirement that both Froude number and Reynolds number need to be held constant between model and prototype. However, when Froude scaling is adopted, it is not possible to assure that Reynolds number is also held constant between model and prototype.

This may be demonstrated as follows. Constancy of Froude number leads to a velocity scale factor given as $k_U = \sqrt{k_L}$. However, the requirement that Reynolds number UL/ν is also preserved between model and prototype implies that the scale factor for viscosity ν would need to correspond to $k_\nu = k_U k_L = k_L^{3/2}$. Most liquids have viscosities that are the same as or lower than water, so that this is impractical if not impossible to achieve under laboratory conditions. Instead, water is used in the laboratory, and the scale effect or disparity with respect to Reynolds number needs to be accepted.

Different approaches to accounting for a disparity in the model Reynolds number have been undertaken, depending on the context. In general, these may involve a simulation of a turbulent flow in artificial ways (such as through trip wires or friction strips), a distinct investigation of Reynolds number effects as a component of a more general modeling study, and/or a combination of laboratory model tests undertaken in conjunction with appropriate numerical modeling. As one example, relevant information on flow separation effects on piles has been developed by the use of a "U-tube" facility in which the flow around a pile section at sufficiently high Reynolds numbers is relied upon, while wave conditions based on Froude modeling are avoided in such tests.

12.4 Laboratory Models in Coastal Engineering

Laboratory models have been used to examine a wide range of topics relevant to coastal engineering, including wave propagation into harbors, tide, tsunami and other long-wave behavior, wave interactions with structures, sediment transport processes, and mixing processes. Typically, Froude scaling is maintained, but it is often not possible to maintain similarity with respect to some of the other relevant dimensionless groups.

12.4.1 Short-Wave Models

Most commonly, studies with wave flows entail Froude scaling, as both gravitational and inertia forces are relevant. These are sometimes referred to as short-wave models to distinguish them from long-wave models (considered below), for which different circumstances apply. This case applies also to open channel or river flows, and to waves generated by marine vessels.

As has been indicated, Froude scaling implies that $k_U = \sqrt{k_L}$. The scale factors for other parameters can then be obtained from various defining relationships. Thus, the scale factor for wave period, k_T, may be expressed in terms of k_L by relying on the defining relationship, velocity = length/time. Since $k_U = \sqrt{k_L}$, this leads to $k_T = k_L/k_U = \sqrt{k_L}$. In coastal engineering applications, freshwater in the laboratory is used in lieu of saltwater in the prototype, so that the scale factor for fluid density is slightly less than 1 ($k_\rho = 0.976$). Dimensionless force coefficients and/or relevant defining relationships can then be used to obtain scale factors for forces and related variables in terms of k_L and k_ρ.

In the above manner, geometric and kinematic similarity and Froude number equivalence give rise to the following scale factor relationships:

$$k_U = \sqrt{k_L}; \quad k_T = \sqrt{k_L}; \quad k_f = 1/\sqrt{k_L}; \quad k_a = 1;$$
$$k_F = k_\rho k_L^3; \quad k_p = k_\rho k_L; \quad k_{F'} = k_\rho k_L^2; \quad k_M = k_\rho k_L^4$$

where L is length, U is velocity, T is period, f is frequency, a is acceleration, F is force, p is pressure, F′ is force per unit length, and M is moment. When freshwater conditions are being modeled or if deemed appropriate, k_ρ would be omitted in the above.

12.4.2 Long-Wave Models

Long-wave models may be used to study long-wave propagation, including tidal flows and tsunami inundation studies, seiche behavior, and estuary or harbor circulation and flushing. Approaches to conducting laboratory tests for long-wave models have been described by Hughes (1993) and by Fischer et al. (1979), with the latter focusing on estuarine and mixing flows.

For long waves such as tides, a key constraint arises because horizontal lengths are generally large (typically several km), so that limitations of laboratory space imply that the horizontal scale must be very small. However, vertical lengths cannot be scaled in the same way, since water depths and wave amplitudes would then be unduly small, such that corresponding measurements would be insufficiently accurate and also invalid because of surface tension scale effects. Since long-wave flows correspond to a wave speed given as $\sqrt{gd}$, where d is the depth, Froude scaling needs to be applied using the water depth and not a horizontal length. Because of this, it is possible to introduce a vertical distortion, such that the vertical length scale factor, denoted k_z, is several times larger than the horizontal length scale factor k_x. Scale factors for related variables can then be obtained in terms of k_x and/or k_z. For example, a characteristic time, such as a tidal period, scales as $k_T = k_x/\sqrt{k_z}$, while a discharge or flow rate Q scales as $k_Q = k_x k_z^{3/2}$. A key consequence of this distortion is that frictional effects along the water boundary do not scale correctly and need to be increased artificially. Typically, this is accomplished by adding metal strips over the seabed in such a manner that a tidal flow in an estuary is reasonably reproduced in the model. Once a model has been calibrated in this way, new situations, such as arising from the influence of engineering works, can be examined.

The extension of long-wave models to mixing studies presents numerous challenges since there is no assurance that relevant features of a mixing flow, taking account of turbulence, dispersion, and stratification will scale correctly. These matters are discussed further by Fischer et al. (1979).

While a number of large long-wave laboratory models have been built and used successfully, the challenges of suitably scaling friction and mixing, coupled with the increased sophistication, reliability, and flexibility of numerical modeling, have led to a greater reliance on numerical models, so that long-wave laboratory modeling is now less common.

12.4.3 Coastal Structures

Laboratory models are used extensively to examine wave and current interactions with coastal structures, including wall-type structures, rubble-mound structures, and floating structures. Model tests have been undertaken to investigate loads on structures, wave overtopping, rubble-mound stability, the motions of floating structures, and the behavior of mooring systems. As stated earlier, a key issue is that Reynolds number does not scale, so that the impact of this disparity needs to be understood and suitably taken into account. Laboratory models are especially useful for wave overtopping, wave impacts, and rubble-mound stability studies.

For floating structures, the dynamic properties and response of the structure, as well as its mooring system, need to be suitably scaled. The various scale factors may readily be established from defining relationships and governing equations in the manner indicated in Section 12.3. Thus, the scale factors for the translational motions (surge, sway, and heave), denoted ξ, are $k_\xi = k_L$, and those for rotational motions (pitch, roll, and yaw), denoted θ, are $k_\theta = 1$. The scale factors for mass m, added mass μ, damping coefficient λ, and stiffness k, need to recognize that freshwater in the model is used to represent saltwater in the prototype, i.e. $k_\rho < 1$. For example, with respect to translational motions, these scale factors are given as $k_m = k_\mu = k_\rho k_L^3$, $k_\lambda = k_\rho k_L^{2.5}$, $k_k = k_\rho k_L^2$. Corresponding expressions may readily be developed with respect to rotational motions and the cross components of m, μ, λ, and k.

12.4.4 Sediment Transport

Laboratory models have been used extensively to study sediment transport behavior. As seen above, it is not possible to assure constancy of the various relevant dimensionless parameters. Even so, these laboratory models can offer significant information. Approaches to conducting laboratory model tests for studying sediment transport have been described by Hughes (1993).

Apart from the need for Froude scaling with respect to wave flows, other scaling requirements relate to sediment size and submerged density, and to the mechanics of the initiation of sediment movement, bedload transport, and suspended load transport.

For the case of suspended load transport, it is the fall velocity and submerged density that are considered to be key parameters. For the cases of incipient motion and bedload transport, the seabed shear stress or friction velocity u_* is a key parameter. Based on the related discussion in Chapter 9 (e.g. see Figure 9.2), the Shields parameter $u_*^2/s'gD$ and the grain size Reynolds number $\mathrm{Re}_* = u_*D/\nu$ are seen to be relevant dimensionless groups. Here, D is the sediment grain size and s' is the submerged specific gravity of the sediment. The Shields parameter is also referred to in this context as a densimetric Froude number.

It is not possible to preserve constancy of the Froude number and the relevant dimensionless groups that involve the various parameters indicated above. Thus, compromises are needed with respect to partial scaling. There are various ways to make such compromises. For instance, the Shields parameter, but not the grain size Reynolds number, is held constant, on the grounds that it may be sufficient to reproduce the Reynolds number flow regime rather than the Reynolds number itself. In some cases, sand is used in the model despite the density and size differences in scaling. More often, the grain size and density are not properly scaled, such that relatively fine, light materials are used. Typically, a model is initially used and adjusted so as to simulate observed sediment transport patterns. Once this has been undertaken, altered conditions, such as the influence of engineering works, can be examined.

Example 12.1 ***Use of Scale Factors***

A model of a pile-restrained floating breakwater is tested in a wave flume using a length scale factor $k_L = 1/20$. While the prototype breakwater is to be located in seawater, freshwater is used in the model. For site conditions corresponding to an incident wave height $H_i = 0.9$ m and period $T = 3$ s, what should be the corresponding wave height and wave period used in the flume? For these waves, measurements in the flume indicate a transmitted wave height $H_t = 1.8$ cm and a maximum force per unit length on the piles of 28 N/m. What will be the transmitted wave height and the maximum force per unit length on the piles that occur in the field?

Solution

Specified parameters

$g = 9.80665\ \text{m/s}^2$
$\rho_p = 1025\ \text{kg/m}^3$
$\rho_m = 1000\ \text{kg/m}^3$
$k_L = 1/20$
$(H_i)_p = 0.9\ \text{m}$
$T_p = 3.0\ \text{s}$
$(H_t)_m = 1.8\ \text{cm}$
$F'_m = 28.0\ \text{N/m}$

***H_i* and *T* in the model**

Based on the relevant scale factor relationships:

$$k_H = k_L = 1/20$$
$$k_T = \sqrt{k_L} = \sqrt{1/20}$$

Therefore:

$$(H_i)_m = (H_i)_p \times k_H = 0.9 \times 1/20\ \text{m} = 4.5\ \text{cm}$$
$$T_m = T_p \times k_T = 3.0 \times \sqrt{1/20}\ \text{s} = 0.67\ \text{s}$$

***H_t* and *F'* in the prototype**

Based on the relevant scale factor relationships:

$$k_\rho = \rho_m/\rho_p = 0.976$$

$$k_{\mathrm{H}} = k_{\mathrm{L}} = 1/20$$

$$k_{\mathrm{F}} = k_{\rho}\ k_{\mathrm{L}}^2 = 1/410$$

Therefore:

$$(H_{\mathrm{t}})_{\mathrm{p}} = (H_{\mathrm{t}})_{\mathrm{m}}/k_{\mathrm{H}} = 1.8/(1/20)\ \mathrm{cm} = 0.36\ \mathrm{m}$$

$$F'_{\mathrm{p}} = F'_{\mathrm{m}}/k_{\mathrm{F}} = 28/(1/410)\mathrm{N/m} = 11.5\ \mathrm{kN/m}$$

12.5 Laboratory Facilities

12.5.1 Kinds of Facilities

Laboratory facilities are used to undertake research into various areas of coastal engineering or to simulate aspects of a coastal engineering project. Figure 12.2 provides photographs of three kinds of facilities that are extensively used – wave flumes, wave basins, and towing tanks.

These facilities are summarized as follows:

- ***Wave flume*** (Figure 12.2a). A long, narrow wave flume is used to simulate a unidirectional wave train and contains a wave generator at one end of the flume and a wave absorber at the other end. Further information is provided in Section 12.5.2.
- ***Wave basin*** (Figure 12.2b). The width of a wave basin is of the order of its length, so that spatially varying waves including oblique waves and multidirectional waves may be simulated. Further information is provided in Section 12.5.3.
- ***Towing tank*** (Figure 12.2c). A towing tank is a long water channel with a moving carriage straddling the tank. It is typically used to test the resistance of ship hulls.

Other facilities include a flow channel for unidirectional (river) flows, which allows for adjustments to various flow and geometric parameters, a "U-tube" facility intended to generate an oscillatory flow past a test section based on resonant oscillations in the facility, and various facilities for examining stratified flows and mixing processes. They can also include an ice tank for examining ocean and river ice interactions with structures, long-wave models that enable the study of tide and tsunami interactions with shorelines and facilities, and sediment transport facilities that enable two-dimensional studies (in a flume) or three-dimensional studies of coastal processes.

Summary information on wave flumes and wave basins is provided below.

12.5.2 Wave Flumes

A wave flume simulates a unidirectional, two-dimensional wave train. As sketched in Figure 12.3, the key features include a wave generator at one end of the flume, a wave absorber at the downwave end, and a test section.

- ***Wave generator***. The wave generator is typically a paddle, a piston, or some other configuration, that is driven sinusoidally to generate periodic waves or has computer-controlled motions to generate random waves or transient waves.

Figure 12.2 Photographs of different kinds of laboratory facilities. (a) Wave flume, (b) wave basin, (c) towing tank. *Source:* Reproduced with permission of National Research Council Canada/Conseil national de recherches Canada.

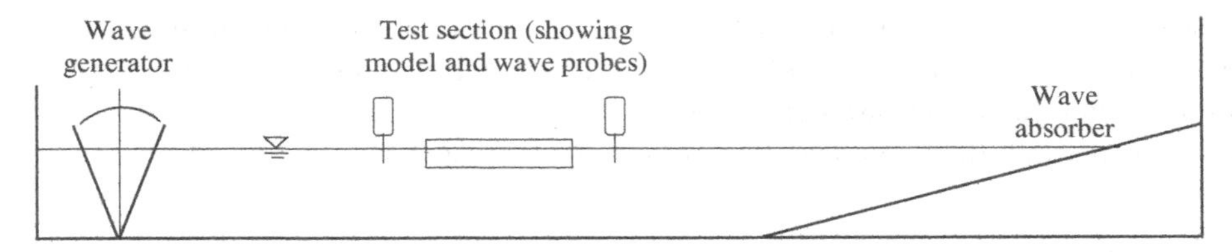

Figure 12.3 Sketch of a wave flume.

- ***Wave absorber***. A wave absorber at the downwave end of the flume is used to minimize wave reflections over the relevant range of wave frequencies so as to avoid undue interference with conditions at the test section. Approaches that are used for wave absorption include an artificial beach (that needs to extend over an appreciable distance in order to be effective), a set of corrugated mesh sheets that seek to limit that horizontal extent of the absorber, and an active wave absorber. The latter refers to a paddle or piston that is programmed so as to oscillate in such a way as to minimize wave reflections.

Alternatively, depending on the context, the beach or other non-transmitting shoreline structure at the downwave end may be the focus of testing.

- ***Test section***. The test section, where measurements are made, is usually located near the center of the flume. This should be away from evanescent waves (i.e. standing waves close to the generator that decay with distance). In some cases, tests are completed before waves reflected from the beach reach the test section.

12.5.3 Wave Basins

Analogous to a wave flume, a wave basin is used to create oblique or three-dimensional waves. Key features of a basin are sketched in Figure 12.4 and include a wave generator, wave absorbers along three sides, and a test area near the center of the basin.

- ***Wave generator***. In its simplest form, the generator is relatively wide and can be relocated as necessary in order to reproduce unidirectional wave trains propagating in different directions. A more sophisticated approach is to employ a segmented wave generator. The use of such a generator relies on the concept whereby sinusoidally oscillating segments, with appropriate phase differences between adjacent segments, produce a wave train propagating obliquely from the generator face. As an extension to this concept, the signals controlling the segment motions may be programmed so as to generate random, short-crested (multidirectional) waves.
- ***Wave absorbers***. The basin walls need to be lined with wave absorbers to minimize wave reflections over the relevant range of wave frequencies in order to avoid undue interference with conditions at the test section. A careful design of these absorbers (e.g. involving corrugated metal sheets) is needed so as to minimize their horizontal extent and thus be effective over the relevant range of wave frequencies. A set of active wave absorbers along the walls may also be used. Unlike the case of a wave flume, these need to be programmed to account for obliquely incident waves and phase differences in the waves as they reach different portions of the walls.
- ***Test section***. The test section is located near the center of the basin where measurements are made.

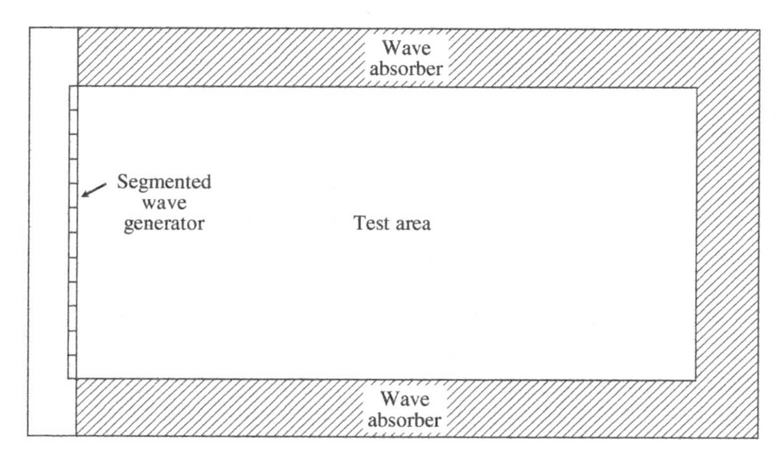

Figure 12.4 Sketch of a wave basin.

12.6 Wave Generation and Measurement

A summary is provided of the processing and analysis used to control single and segmented wave generators, and of the processing and analysis of measurements from instruments such as wave probes, accelerometers, and pressure transducers.

12.6.1 Wave Generator Control

A generator in a flume may be controlled by a mechanical control system to undergo periodic motions, thus creating regular waves with different frequencies and heights. Information on developing the frequency-dependent transfer function that relates wave height to generator stroke for a given generator configuration was provided in Section 6.5.

For irregular waves in a flume, the signal controlling the generator motion may be programmed so as to produce either random waves with a specific wave spectrum, breaking waves, or episodic waves. The programming of such a signal takes account of the frequency-dependent transfer function between generator amplitude and wave amplitude.

For a segmented generator in a wave basin, the programming needs to take account of phase differences between adjacent segments so as to develop oblique waves with a prescribed direction and then a superposition of these component signals so as to develop a specified directional spectrum.

12.6.2 Instrumentation and Measurement Techniques

Measurements in the laboratory may relate to water surface elevation, wave runup, pressure, forces, moments, flow displacements, velocities and accelerations, fluid densities, pollutant concentrations, sediment properties, and sediment transport. Most of these measurements rely on various kinds of instrumentation. Techniques are available to analyze signals from a wave probe or multiple wave probes or other measurements such as those from accelerometers or pressure transducers. These may yield amplitudes, frequencies, one-dimensional wave or response spectra, and directional wave spectra. Hughes (1993), for example, provides detailed information on different kinds of instruments, different kinds of measurements that are taken, and the analyses of measurement signals that are needed.

12.7 Field Measurements

Finally, brief mention is made of field measurement programs that may be undertaken to obtain measurements of a wide range of parameters of interest in coastal engineering. In general, these may relate to waves, water flows and water properties, sediments and bathymetry, and the design and performance of coastal structures. Field measurements may be undertaken in order to develop databases of ocean-related measurements (e.g. tidal records and wind station records) for research or for project-specific purposes. Limitations with respect to project-specific measurements include their high costs relative to a project's budget and the limited duration over which measurements are made, which usually precludes a capture of extreme design conditions. One approach is to rely on project-specific field measurements, such as wave buoy measurements, in order to calibrate or validate numerical models, which can in turn be used to develop design conditions relevant to a particular project.

A summary of measurement methods for a range of relevant parameters is indicated below.

Waves. Wave heights, wave periods, wave height/period distributions, and one-dimensional wave spectra may be obtained from the surface elevation record at a single location. The surface elevation may be measured by a buoy containing an accelerometer, and various forms of wave gauge, wave staff, or wave sensor. The latter are usually based on acoustic or pressure measurements, including a bottom-mounted pressure gauge and an acoustic surface-tracking sensor. Wave directional spectra may be developed from multiple output signals obtained from point array gauges (several wave gauges arranged in close proximity), a heave-pitch-roll buoy, or an instrumentation package measuring surface elevation and horizontal velocity components. Significant wave heights over a broader area of the ocean may be measured by altimeters located in satellites.

Tides. Long wave water levels, including tide levels, may be measured by tide gauges or related instruments that electronically or mechanically filter out higher frequency components.

Currents. Current speeds and directions may be measured by a wide range of current meters. These include propellor, electromagnetic and ultrasonic current meters, an acoustic Doppler current profiler (ADCP) that measures currents across a water column, and optical measurements of the paths of floats or dye that follow the water motion.

Water properties. Water properties that are measured include water density, salinity, and turbidity. The measurements may need to take account of their vertical and horizontal variations in a given area.

Wind. Wind measurement programs based on anemometer measurements at many wind stations over extended durations are well established in many countries.

Sediments. Field measurements relating to sediments and morphology include the measurement of sediment properties (including grain size distributions obtained from sieve analyses), beach profiles from surveying methods, suspended load concentrations, and sediment transport rates (including a reliance on bedload traps). Surveying methods include total station topographic surveying, or drone and lidar technology. (Lidar stands for Light Detection and Ranging and uses a pulsed laser to measure distances.) An orthophotograph (an aerial photograph geometrically corrected such that the scale is uniform) taken from a drone survey or through other methods may be used to obtain various measurements relating to a shoreline.

Bathymetry. Bathymetry may be measured using depth sounders, including those reliant on sonar and lidar techniques, and satellite altimetry.

Coastal structures. Instrumentation relating to coastal structures is used for the measurement of pressures, wave runup, wave overtopping, wave behavior adjacent to a structure (including reflected and transmitted waves), the motions of a floating structure, and the stability of armor units. Motions may be measured by an inertial measurement unit (IMU), an optical tracking system, or a suitable instrumentation package. Armor unit displacements can be assessed by surveys of armor units and stereophotogrammetry using aerial photographs and side-scan sonar to investigate the condition of underwater armor units. The loads and motions of mooring systems and anchors can be measured in various ways.

Problems

[Where relevant, you should rely on the values of physical constants provided in Appendix C.]

Assignments

12.1 In a model test in a wave flume, a wave period of 1.5 s in the flume corresponds to a wave period of 6.0 s at full scale. If the force on a pile in the model is measured to be 100 N, what will be the corresponding force for the full-scale situation? What is the ratio of Reynolds numbers in the model relative to the prototype? Assume that freshwater is used in the model, while saltwater is present in the prototype.

12.2 A wave flume is to be used to carry out model tests to investigate wave forces on a submerged cylindrical caisson with a diameter of 8 m and a height of 6 m in water of depth 20 m. The design wave height is 5 m and the wave period is 10 s. The flume is 3.6 m wide and 4.6 m deep and can produce waves within the frequency range 0.25–1.0 Hz. Recognizing that the model caisson diameter should be no more than about a third of the flume width, select suitable dimensions for the model caisson. Determine the water depth, wave height, and wave frequency to be used. Assuming that freshwater is used in the model, while saltwater is present in the prototype, determine the resulting scale factors for force and overturning moment.

Written Assignment – Laboratory Report

12.3 You are required to conduct a laboratory model test in a wave flume in order to investigate the performance of a section of a novel floating breakwater restrained by piles. The objectives of the study are to determine wave transmission coefficients, breakwater motions, and maximum loads on the piles per unit length of breakwater, all for a range of incident wave conditions. The proposed breakwater has a beam of 6 m and a draft of 1.2 m. It will be located in water depths ranging from 3 to 8 m, and it may be subjected to waves with heights up to 1.5 m and periods up to 3.5 s. Presuming the use of a wave flume with suitable characteristics and your successful completion of a hypothetical test program, develop a summary report (using bullet format to the extent possible) describing the study. The report should include the following components: a selection of breakwater model dimensions and parameters; a description of the instrumentation to be used and measurements to be made; a summary outline of the tests that have been undertaken; a discussion of constraints or limitations in undertaking the tests; and a description of the test results, relying on the use of dimensionless groups to the extent possible.

H_d	design wave height
H_i	incident wave height
H_m	maximum wave height
H_o	deep-water wave height
H_r	root-mean-square wave height; reflected wave height
H_s	significant wave height
H_{sb}	significant wave height at breaking line
H_{st}	significant wave height near structure toe
h	structure crest elevation above mean water level; draft
i	imaginary i
K	Keulegan–Carpenter number; longshore transport coefficient
K_d	diffraction coefficient
K_D	armor stability coefficient
K_e	energy dissipation/extraction ratio
K_r	refraction coefficient; reflection coefficient
K_s	shoaling coefficient
K_t	transmission coefficient
k	wave number; stiffness;
k_s	sediment roughness parameter
k_x	scale factor of variable x
L	wave length; design life
L_o	deep-water wave length basin length; harbor length; estuary length
M	overturning moment; mass of armor unit; mass of pollutant; specific momentum flux
M'	overturning moment per unit width
M'_s	hydrostatic moment per unit width
m	beach slope; mass

m_a	added mass
m (subscript)	value of variable in model
N	number of waves
n	Manning n; porosity
P	cumulative probability; wave energy flux
P_ℓ	longshore wave energy flux
p	pressure (relative to atmospheric pressure); probability density
p_a	atmospheric pressure
p_b	maximum breaking wave pressure
p_e	pressure at eye of hurricane
p_o	pressure at seabed
p_t	pressure at top of wall
p (subscript)	value of variable in prototype
Q	exceedance probability; longshore transport rate; specific mass flux
Q_f	freshwater inflow rate
R	wave runup; tidal exchange ratio
$R_{2\%}$	wave runup of highest 2% of waves
$\overline{\mathrm{R}}$	average wave runup
Re	Reynolds number
Re_w	$= au_o/\nu$, wave Reynolds number
Re_*	$= u_* D/\nu$, grain size Reynolds number
R_m	maximum wave runup
r	radial coordinate
r_i	recording interval
r_y	radius of gyration
S	radiation stress; damage level; Strouhal number; salinity
$S(f)$	spectral density
$S(f,\theta)$	directional wave spectrum
$\overline{S}$	mean salinity

S_e	average salinity of water exiting an estuary
S_m	maximum scour depth
S_o	ocean water salinity
s	vertical coordinate measured upwards from the seabed;
	wind or wave setup
s'	submerged specific gravity
$s(x, t)$	offshore shift in beach profile
s_o	wind or wave setup at shoreline
T	wave period
T_f	wave period in the presence of a current relative to a fixed reference frame
T_F	transfer function;
	flushing time
T_h	heave natural period
T_n	period of nth-mode of harbor oscillations
T_p	peak period
T_R	return period
T_r	roll natural period
T_z	zero-crossing period
t	time
t_m	time of maximum force
U	steady velocity;
	depth-averaged velocity;
	current speed;
	wind speed
$\hat{U}(\tau)$	largest wind speed averaged over duration τ
U_1	1-hr average wind speed
U_m	current magnitude
u	horizontal component of fluid velocity
u_o	velocity amplitude at seabed
u_*	friction velocity
$\dot{u}$	horizontal component of fluid acceleration
u_m	maximum oscillatory velocity
v	velocity
V	vessel speed;
	estuary volume at high tide
$\overline{V}$	average longshore current
V_f	freshwater volume in an estuary

w	vertical component of fluid velocity; gap below barrier; jet thickness; fall velocity
$\dot{w}$	vertical component of fluid acceleration
w_{ij}	wave scatter diagram entries
X	time-varying force/moment given by the Morison equation; generator displacement amplitude
X_{m}	maximum of X
x	horizontal coordinate in wave direction; abscissa used in EVA plot
$\dot{x}$	velocity
$\ddot{x}$	acceleration
x_{m}	maximum of x
y	horizontal coordinate transverse to wave direction; ordinate used in EVA plot
z	vertical coordinate measured upwards from the SWL; elevation
z_{B}	z value of center of buoyancy
z_{G}	z value of center of gravity
α	wave direction; wind direction; current direction
α_{b}	wave direction at breaking
α_{o}	deep-water wave direction
β	beach slope angle; wave direction used in Goda formulation
γ_{b}	breaker index
Δ	MWL above SWL used in Miche-Rundgren/Sainflou method; tidal prism
ΔS	end-year minus current year median RSLR
δ	phase angle of force; phase angle of tidal constituents
ε	phase angle; amplitude perturbation parameter;
ζ	vertical component of fluid displacement; damping ratio
η	free surface elevation above SWL

η_b	breaking wave crest elevation
η_c	crest elevation
θ	angular coordinate;
	wave direction
$\overline{\theta}$	mean wave direction
θ_o	incident wave direction for diffraction
κ	effective drag coefficient
λ	damping coefficient
μ	added mass;
	relative depth perturbation parameter
μ_H	mode of wave height
ν	kinematic viscosity
ξ	horizontal component of fluid displacement;
	surf similarity parameter
π	dimensionless group
ρ	water density
ρ_f	freshwater density
ρ_s	density of sediment, rock, or armor
σ_η	standard deviation of free surface elevation
τ	shear stress;
	duration;
τ_c	threshold duration
τ_b	current-induced shear stress at seabed
τ_s	shear stress on water surface
τ_w	shear stress at seabed
ϕ	velocity potential
ϕ_s	scattered wave potential
ϕ_w	incident wave potential
ψ	stream function
ω	wave angular frequency
ω_f	wave angular frequency in the presence of a current relative to a fixed reference frame

C

Physical Constants

This appendix provides values of physical constants that are used most frequently in coastal engineering. The values of density, viscosity, and salinity given are approximate, intended for ready application to coastal engineering calculations. More specific values should be used in particular circumstances. Thus, material density varies with the specific material that is used, while water density, viscosity, and salinity depend on various factors such as temperature and location.

Gravitational constant, g	$9.80665\ \mathrm{m/s^2}$
Standard atmospheric pressure, p_a	1013.25 mbar or 101.325 kPa
Density of seawater, ρ	$1025\ \mathrm{kg/m^3}$
Density of freshwater, ρ_f	$1000\ \mathrm{kg/m^3}$
Density of sediment or rock, ρ_s	$2650\ \mathrm{kg/m^3}$
Density of concrete armor units, ρ_s	$2400\ \mathrm{kg/m^3}$
Dry bulk density of sand	$1600\ \mathrm{kg/m^3}$
Saturated bulk density of sand	$2000\ \mathrm{kg/m^3}$
Kinematic viscosity of water, ν	$1.0 \times 10^{-6}\ \mathrm{m^2/s}$
Salinity of seawater, S_o	35 g/kg

An Introduction to Coastal Engineering, First Edition. Michael Isaacson.

Companion website: www.wiley.com/go/coastalengineering

References

American Society of Civil Engineers (ASCE) (2006). *Flood Resistant Design and Construction*. ASCE Standard ASCE/SEI 24-14.

American Society of Civil Engineers (ASCE) (2012). *Planning and Design Guidelines for Small Craft Harbors*, 3rd edition. ASCE Manuals and Reports on Engineering Practice No. 50.

American Society of Civil Engineers (ASCE) (2022). *Minimum Design Loads and Associated Criteria for Building and Other Structures*. ASCE Standard ASCE/SEI 7-22.

Battjes, J.A. (1970). *Long-term wave height distribution at seven stations around the British Isles*. National Institute of Oceanography, UK, Report No. A44.

Berkhoff, J.C.W. (1972). Computation of combined refraction-diffraction. In: *Proceedings of the 13th International Conference on Coastal Engineering, Vancouver, Canada (July 10–14, 1972)*, 471–490. ASCE.

British Standards Institution (BSI) (1991). *Maritime Structures – Guide to the Design and Construction of Breakwaters*. Standard BS 6349-7: 1991.

Bruun, P. (1981). *Port Engineering*, 3rd edition. Gulf Publishing Co.

Coastal Engineering Manual (2002). U.S. Army Corps of Engineers. 2002 (with subsequent updates).

Construction Industry Research and Information Association (CIRIA) (2010). *Beach Management Manual*, 2nd edition. Construction Industry Research and Information Association (CIRIA) C685.

Construction Industry Research and Information Association (CIRIA) (2020). *Groynes in Coastal Engineering. Guide to Design, Monitoring, Maintenance of Narrow Footprint Groynes*. Construction Industry Research and Information Association (CIRIA).

Cox, A.T. and Swail, V.R. (2001). A global wave hindcast over the period 1958–1997: validation and climate assessment. *Journal of Geophysical Research* 106 (C2): 2313–2329.

Dalrymple, R.A. (1988). Model for refraction of water waves. *Journal of Waterway, Port, Coastal, and Ocean Engineering, ASCE* 114 (4): 423–435.

Dean, R.G. and Dalrymple, R.A. (1991). *Water Wave Mechanics for Engineers and Scientists*. World Scientific.

EurOtop (2018). *Manual on Wave Overtopping of Sea Defenses and Related Structures. An Overtopping Manual Largely Based on European Research, but for Worldwide Application*, 2nd edition. overtopping-manual.com.

Federal Emergency Management Agency (FEMA) (2009). *Recommended Residential Construction for Coastal Areas. FEMA P-550*, 2nd edition.

Federal Emergency Management Agency (FEMA) (2011). *Coastal Construction Manual: Principles and Practices of Planning, Siting, Designing, Constructing and Maintaining Residential Buildings in Coastal Areas*, 4th edition.

An Introduction to Coastal Engineering, First Edition. Michael Isaacson.

Companion website: www.wiley.com/go/coastalengineering

Federal Emergency Management Agency (FEMA) (2023). *Guidance for Flood Risk Analysis and Mapping. Coastal Overland Wave Propagation*. Guidance Document No. 41.

Fenton, J.D. (1988). The numerical solution of steady water wave problems. *Computers and Geosciences* 14: 357–368.

Ferguson, R.I. and Church, M. (2004). A simple universal equation for grain settling velocity. *Journal of Sedimentary Research* 74 (6): 933–937.

Fine, I.V., Cherniawsky, J.Y., Rabinovich, A.B., and Stephenson, F. (2009). Numerical modeling and observations of tsunami waves in Alberni Inlet and Barkley Sound, British Columbia. *Pure and Applied Geophysics* 165: 2019–2044.

Fischer, H.B., List, E.J., Koh, R.C.Y. et al. (1979). *Mixing in Inland and Coastal Waters*. Academic Press.

Fournier, C.P., Mulcahy, M.W., Chow, K.A., and Sayao, O.J. (1992). Wave agitation criteria for fishing harbours in Atlantic Canada. In: *Proceedings of the 23rd International Conference on Coastal Engineering Conference, Venice, Italy (October 4–9, 1992)*, 3230–3243. ASCE.

Hughes, S.A. (1993). *Physical Models and Laboratory Techniques in Coastal Engineering*. World Scientific.

Imberger, J. (2012). *Flow Processes, Scaling, Equations of Motion, and Solutions to Environmental Flows*. Elsevier.

Intergovernmental Panel on Climate Change (IPCC) (2021). *Climate Change 2021: The Physical Science Basis. Contribution of Working Group I to the Sixth Assessment Report of the Intergovernmental Panel on Climate Change*. Cambridge University Press. [Note: Chapter 9, Ocean, Cryosphere and Sea Level Change, and Chapter 11, Weather and Climate Extreme Events in a Changing Climate, are of particular relevance to coastal engineers.]

International Organization for Standardization (ISO) (2007). *Actions from Waves and Currents on Coastal Structures*. ISO Standard 21650.

Ippen, A.T. (ed.) (1966). *Estuary and Coastline Hydrodynamics*. McGraw-Hill.

Isaacson, M. and Mercer, A.G. (1982). The response of small craft to wave action. In: *Proceedings of the 18th International Conference on Coastal Engineering, Cape Town, South Africa (November 14–19, 1982)*, 2723–2742. ASCE.

Kamphuis, J.W. (1975). Friction factor under oscillatory waves. *Journal of the Waterways, Harbors and Coastal Engineering Division, ASCE* 101 (WW2): 135–144.

Kamphuis, J.W. (2020). *Introduction to Coastal Engineering and Management*, 3rd edition. World Scientific.

Kraus, N.C. (ed.) (1996). *History and Heritage of Coastal Engineering*. American Society of Civil Engineers.

Kriebel, D.L. and Bollmann, C.A. (1996). Wave transmission past vertical wave barriers. In: *Proceedings of the 25th Conference on Coastal Engineering, American Society of Civil Engineers, Orlando, FL. (September 2–6, 1996)*, 2470–2483. ASCE.

Limber, P.W., Barnard, P.L., Vitousek, S., and Erikson, L.H. (2018). A model ensemble for projecting multidecadal coastal cliff retreat during the 21st century. *Journal of Geophysical Research: Earth Surface* 123: 1566–1589.

Peregrine, D.H. (1967). Long waves on a beach. *Journal of Fluid Mechanics* 27 (4): 815–827.

Reeve, D., Chadwick, A., and Fleming, C. (2018). *Coastal Engineering: Processes, Theory and Design Practice*, 3rd edition. CRC Press.

Robertson, B., Hall, K., Zytner, R., and Nistor, I. (2013). Breaking waves: review of characteristic relationships. *Coastal Engineering Journal* 55 (1): 1350002-1–1350002-40.

Robin, C.M.I., Craymer, M., Ferland, R. et al. (2020). *NAV83v70VG: A new national crustal velocity model for Canada*. Geomatics Canada Open File 0062.

Rock Manual (2007). 2nd edition. Construction Industry Research and Information Association (CIRIA).

Rubin, H. and Atkinson, J. (2001). *Environmental Fluid Mechanics*. Marcel Dekker Re-published 2019 by CRC Press.

Sarpkaya, T. and Isaacson, M. (1981). *Mechanics of Wave Forces on Offshore Structures*. Van Nostrand Reinhold.

Sawaragi, T. (2011). *Coastal Engineering – Waves, Beaches, Wave-Structure Interactions*. Elsevier.

Shen, H.H., Cheng, A.H.D., Wan, K.-H. et al. (ed.) (2002). *Environmental Fluid Mechanics: Theories and Applications*. ASCE.

Shore Protection Manual (1984). 4th edition. U.S. Army Corps of Engineers.

Sleath, J.F.A. (1984). *Sea Bed Mechanics*. Wiley-Interscience.

Sorensen, R.M. (2006). *Basic Coastal Engineering*, 3rd edition. Springer.

Sumer, B.M. and Fredsoe, J. (2002). *The Mechanics of Scour in the Marine Environment*. World Scientific.

Tobiasson, B.O. and Kollmeyer, R.C. (1991). *Marinas and Small Craft Harbors*. Van Nostrand Reinhold.

Van der Meer, J. and Sigurdarson, S. (2017). *Design and Construction of Berm Breakwaters*. World Scientific.

Weigel, R.L. (1964). *Oceanographical Engineering*. Prentice-Hall.

l

m

n

o

p

r

t